T0218322

The Beauty of Mathematics in Computer Science

The Beauty of Mathematics in Computer Science

Jun Wu

Translated from the Chinese edition by

Rachel Wu and Yuxi Candice Wang

CRC Press
Taylor & Francis Group
Boca Raton London New York

CRC Press is an imprint of the
Taylor & Francis Group, an **informa** business

A CHAPMAN & HALL BOOK

CRC Press
Taylor & Francis Group
6000 Broken Sound Parkway NW, Suite 300
Boca Raton, FL 33487-2742

International Standard Book Number-13: 978-1-138-04960-4 (Paperback)
978-1-138-04967-3 (Hardback)

Library of Congress Cataloging-in-Publication Data

Names: Wu, Jun, 1967- author.
Title: The beauty of mathematics in computer science / Jun Wu.
Description: Boca Raton, FL : Taylor & Francis Group, 2019.
Identifiers: LCCN 2018035719| ISBN 9781138049604 paperback | ISBN 9781138049673 hardback
Subjects: LCSH: Computer science--Mathematics. | Machine learning.
Classification: LCC QA76.9.M35 W84 2019 | DDC 004.01/51--dc23
LC record available at https://lccn.loc.gov/2018035719

**Visit the Taylor & Francis Web site at
http://www.taylorandfrancis.com**

**and the CRC Press Web site at
http://www.crcpress.com**

Contents

Foreword

A few years ago, I wrote the forewords for Dr. Jun Wu's Chinese editions of *On Top of Tides* and *The Beauty of Mathematics*. Since, I'm happy to learn that *The Beauty of Mathematics* was awarded the prestigious Wenjin Prize.

The Beauty of Mathematics in Computer Science, with its new name in the English edition, originates from a series of Google China blog articles by Google's senior staff research scientist, Jun Wu. Initially, the blog's editors were afraid that the subject matter would bore readers—that it was too abstract and esoteric. That fear was quickly dispelled. Through vivid and compelling language, *The Beauty of Mathematics in Computer Science* connects the history of mathematics to the emergence of its practical applications. Wu systematically introduces the reader to important mathematical bases from which modern science and technology arise. Complex ideas are made accessible to a wide audience, especially those interested in science and technology.

As I wrote in the preface to *On Top of Tides*, Wu is one of a rare kind, with both strong narrative abilities and deep insight into the development of modern technology. Among the researchers and engineers I know, Wu stands in a unique position to share effectively his knowledge with the general public. In *The Beauty of Mathematics in Computer Science*, Wu proves this point once again. From years of accumulated knowledge, Wu excels in his grasp of mathematics and information processing, introduced here as professional disciplines ranging from speech recognition and natural language processing to information search. From the origins of digital computing, to the mathematics behind search engines, and clever mathematical applications to search, Wu brings the essence of mathematics to life. In his writing, mathematics is not a bag of boring, abstruse symbols; rather, it drives fascinating technologies in our everyday lives. Indeed, through Wu's tale, the reader will inevitably discover the hidden beauty of mathematics.

Galileo once said that "mathematics is the language with which God has written the universe." Along the same line of thought, Einstein wrote, in an obituary to Amalie Emmy Noether:

> Pure mathematics is, in its way, the poetry of logical ideas ... In this effort toward logical beauty spiritual formulas are discovered necessary for the deeper penetration into the laws of nature.

From years of personal study in information processing and speech recognition, I deeply appreciate the fundamental role mathematics plays in all fields of science.

In the fifth century AD, Greek philosopher Proclus Diadochus quipped that "wherever there is number, there is beauty." Today, I sincerely recommend *The Beauty of Mathematics in Computer Science* to any friend interested in the natural sciences, scientific research, or life in general. Whether you study liberal arts or engineering, Wu's exposition on mathematics will help you appreciate the elegance and sublimity of our universe. The value of this book lies in both the author's familiarity with the subject matter and his active role in developing such technologies as a career. Wu not only explains why simple mathematical models can solve complex engineering problems, but also reveals the thought processes behind his colleagues' and his own work. Without experience in application, most scholars of pure mathematics cannot achieve the latter point.

From the original Google China blog articles to the publication of *The Beauty of Mathematics in Computer Science*, Wu has spent much time and effort on this book. During his spare time from work, he rigorously rewrote most of the articles, so that an ordinary reader could understand and enjoy the material, but an expert would still learn much from its depth. Since the first version, Wu has encapsulated two more years of research at Google into two new chapters. In this edition, I hope that the reader will further appreciate the beauty of mathematics.

Sometimes, I find that today's world is under a lot of pressure to be practical, and in the process, has lost some of its curiosity about the natural world. In this respect, Wu's book is an excellent remedy. I very much hope that in the future, Wu will continue to write such books, elucidating complex ideas in simple terms. These ideas are the best gifts he could bestow upon society and the younger generation.

Kai-Fu Lee

Preface

The word "mathematics" stems from the Greek word, $\mu\acute{\alpha}\theta\eta\mu\alpha$, which means "wisdom gained from learning." In that sense, early mathematics encompassed a wider field of study and related more closely to people's everyday lives.

Early mathematics was less mysterious and more practical. Like any field, mathematics undergoes constant change, becoming deeper and more theoretical in the process. In fact, the evolution of mathematics corresponds to the constant abstraction of concrete experiences and their effects on our lives. Today, several millennia have stripped mathematics down to numbers, symbols, equations, and theorems—quite detached from our everyday lives. People might use arithmetic to calculate the amount to tip at a restaurant, but beyond those basics, many see little value in mathematics, especially pure mathematics. After graduation, most college students will never encounter advanced mathematics again, so after a few years, they forget most of what they have learned. As such, many question the point of learning mathematics in the first place.

Even worse, many mathematicians have difficulty making a living from pure mathematics, both in the United States and China. A common stereotype portrays all mathematicians as hopeless "nerds," with thick-rimmed glasses and poor social skills. To the layperson, why doom yourself to the socially unattractive label of "mathematician"? Thus, neither the abstruse numbers and symbols, nor the peculiar folk who study them, evoke common signs of beauty.

On the contrary, mathematics is far more prevalent in our lives than we may think. Initially, we might consider only scientists as direct consumers of mathematics—that is, of course studying atomic energies or aerodynamics would require mathematical knowledge! If we look deeper, though, technologies we use every day are all constructed from mathematical building blocks. When you ask Siri for today's weather, mathematical gears whir into action. As a researcher for over twenty years, I have often marveled at the ways mathematics can be applied to practical problems, which are solved in a way I can only describe as magic. Therefore, I hope to share some of this magic with you.

In ancient times, the most important skill people developed, besides knowledge of the natural world, was to exchange ideas, or broadly, to communicate. Accordingly, I have selected *communication* as the starting point of this book. Communication is an excellent field to illustrate the beauty of mathematics, for not only does it abundantly adopt mathematics as a tool, but it also ties closely to our everyday lives.

Since the Industrial Revolution, communication has occupied a large percentage of time in people's lives. Moreover, after the advent of electricity, communication has both closed the distance between people and accelerated the world's economic growth. Today, it is commonplace to spend copious amounts of time lounging in front of a TV or watching Netflix, browsing the latest social media or posting photos from a smart phone. These are all offshoots of modern communication. Even activities that traditionally involved physically traveling somewhere, like shopping, have been overtaken by e-commerce sites—again, modern communication technologies. From century-old inventions, like Morse's telegraph and Bell's phone, to today's mobile phones and Internet, all modern implementations of communication have adhered to information theory principles, which are rooted in mathematics. If we search further back, even the development of language and writing were based on mathematical foundations.

Consider some of our everyday technologies: omnipotent web search, locating the tastiest restaurants and your local DMV; speech recognition on a smart phone, setting a reminder for laundry; or even online translation services, preferably not for your child's foreign language homework. To an everyday user, it is not immediately evident that mathematics drives all these seemingly magical features. Upon further inspection, however, these diverse applications can all be described by simple mathematical models. When engineers find the appropriate mathematical tool for some hairy problem, they will often bask in the elegance of their solution. For instance, although there are hundreds of human languages, from English to Swahili, the underlying mathematical models for translating them are the same, or nearly so. In this simplicity lies beauty. This book will introduce some of these models and demonstrate how they process information, especially to bring about the technological products we use today.

There is often an aura of mystery about mathematics, but its essence is uncomplicated and straightforward. English philosopher, Francis Bacon, once quipped, "Virtue is like a rich stone, best plain set." Mathematics is precisely this type of virtue. Thus throughout this book, I will attempt to portray that simplicity is beauty.

Finally, I provide a brief explanation for the book's extensive treatment of natural language processing ideas and in particular, its experts. These world-class scholars hail from a diverse set of nationalities or backgrounds, but they share a common love for mathematics and apply its methods towards practical problems. By recounting their lives and daily work, I hope the reader can better understand these individuals—understand their ordinariness and excellence; grasp their reasons for success; and most of all, sense that those who discover the beauty in mathematics live more fulfilling lives.

Jun Wu

Acknowledgments

I would like to thank my wife, Yan Zhang, for her longtime support and generous help in my career, and my father Jiaqing and my mother Xiuzhen Zhu, who brought me to the world of mathematics when I was very young.

I would like to thank many of advisors, colleagues, and friends, who gave me their generous and wise advice and suggestions during my career, especially to Prof. Zuoying Wang of Tsinghua University, Prof. Sanjeev Khudanpur, Prof. Fred Jelinek, and Prof. David Yarowsky of Johns Hopkins University, Dr. Eric Brill of eBay, Prof. Michael Collins of Columbia University, Dr. Amit Singhal, ex-senior VP of Google, Mr. Matt Cutts, Dr. Peter Norvig, Dr. Kai-Fu Lee, Dr. Franz Och, Ms. Dandan Wu, Ms. Jin Cui, and Dr. Pei Cao of Google. I am also grateful to Prof. Wing H. Wong of Stanford University and Mr. John Kimmel of Chapman & Hall, who helped me to publish this book.

Special thanks to Rachel Wu and Yuxi Wang for translating the book from Chinese, and Ms. Teena Lawrence and Ms. Michele Dimont, the project managers of the book.

Chapter 1

Words and languages, numbers and information

Words and numbers are close kin; both building blocks of information, they are as intricately linked as elements of nature. Language and mathematics are thereby connected by their shared purpose of recording and transmitting information. However, people only realized this commonality after Claude E. Shannon proposed the field of information theory, seven decades ago.

Since ancient times, the development of mathematics has been closely tied to the ever-growing human understanding of the world. Many fields—including astronomy, engineering, economics, physics, and even biology—depended on mathematics, and in turn, provided new grounds for mathematics to advance. In the past, however, it was quite unheard of for linguistics to draw from mathematics, or vice versa. Many famous mathematicians were also physicists or astronomers, but very few were also linguists. Until recently, the two fields appeared incompatible.

Most of this book tells the story of the past half century or so, but in this chapter, we will venture back to ancient history, when writing and numbers were first invented.

1.1   Information

Before our *Homo sapiens* ancestors developed technologies or started looking like modern humans, they could convey information to each other. Just as zoo animals make unintelligible animal noises, early humans made unintelligible "human" sounds. Initially, maybe the sounds had little meaning beyond exercising available muscles, but gradually, they began to carry messages. For example, some series of grunts may signify, "there's a bear!". To which a companion may grunt "yuh" in acknowledgment, or another series of sounds that signify, "let's go pelt it with rocks." See Figures 1.1 and 1.2.

In principle, there is little difference between these primordial grunts and the latest methods of information transmission, reception, and response. We will fill in the details of communication models in later chapters, but note here that simple models capture both ancient and modern communication.

Early humans did not understand enough about the world to communicate much, so they had no need for language or numbers. However, as humankind

FIGURE 1.1: Earliest forms of communication for humankind.

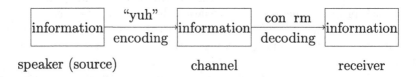

speaker (source)   channel   receiver

FIGURE 1.2: Same communication model beneath primordial grunts and modern information transfer.

progressed and civilization developed, there was a growing need to express more and more information. A few simple noises could no longer fulfill humans' communication needs, and thus language was invented. Stories of daily life, which can be considered a specific type of data, were actually the most valuable artifacts from that time. Transmitted through oral tradition, these stories were passed down the generations, through each cycle of offspring. When humans began to accumulate material goods or food surpluses, the concepts of "more" and "less" emerged. In these days, counting had not yet been invented, since there was no need to count.

1.2   Words and numbers

Our ancestors quickly learned more about their world, and their language became richer, more abstract in the process. Elements that were often described, including objects, numbers, and actions, were abstracted into words of their own, precursors to the vast vocabularies of today. When language and vocabulary grew to a certain size, they could no longer fit inside a single human brain—just as no one today can remember all the wisdom of mankind. A need for efficient recording of information arose, and its solution was writing.

Archeologists today can verify when writing (including numbers) first appeared. Many readers of *On Top of Tides* have asked me why that book primarily discusses companies in the United States. The reason is simple: the past hundred years of technological revolutions have been set almost completely there. Likewise, to study the information revolutions of 5,000 to 10,000 years ago, we must return to the continent of Africa, where our human ancestors first walked out, the cradle of human civilization.

A few thousand years before the oldest (discovered) Chinese oracle bones were carved, the Nile River Valley was already nurturing an advanced civilization. Ancient Egyptians were not only excellent farmers and architects, but they were also the earliest inventors of ideographs.* These were the famed hieroglyphics. Ancient Egyptian left many hieroglyphical scrolls describing their lives and religions. One of the most famous hieroglyphical scrolls is "Book of the Dead (Papyrus of Ani.)", which resides permanently in the British Museum. The scroll consists of over 20 meters of painted papyrus, with over 60 paintings and pictographs. This cultural relic portrays an all-encompassing record of Egyptian civilization, 3,300-3,400 ago.

In early days of Egyptian civilization, the number of existing hieroglyphics corresponded directly to the amount of information to be documented. The earliest engravings of hieroglyphics, dating back to the 32nd century BC, utilized only around 500 characters. By the fifth century BC (the classical Greco-Roman era), this number had increased 5,000 characters, approximately the number of commonly used Chinese characters. However, as civilization continued to develop, the number of hieroglyphics did not increase with the production of information. There is a finite limit to the number of characters any one person

FIGURE 1.3: Rosetta Stone.

---

*Images that represent objects or ideas.

can remember, so instead of inventing more characters, ancient civilizations began to generalize and categorize concepts. For example, the ancient Chinese ideograph for "day" represented both the sun itself, as well as the time between sunrise and sunset. In ancient Egyptian hieroglyphics, a single symbol could also convey multiple meanings.

This idea of clustering ideas into single characters is similar to today's concept of "clustering" in natural language processing or machine learning. Ancient Egyptians required thousands of years to consolidate multiple meanings into a single word. Today, computers may take several hours, or even minutes, to accomplish the same task.

When a single word can take on many meanings, ambiguities inevitably emerge. Given different environments, a word can dance from one meaning to another. Disambiguation, or determining the specific meaning, has not changed from traditional linguists to modern computers: we must examine the context. In most cases, the context will tell us the answer, but of course, there are always outliers. Consider your favorite religious text. Many scholars, students, or other theologians have expounded on these texts, but there exists no interpretation without controversy. Each scholar will use his or her own understanding to eliminate ambiguity, but none have flawlessly succeeded thus far, otherwise many a bloody war would have been averted. Thus is the inconclusive nature of human language. In our case, the situation is similar. Even the most successful probabilistic models will fail at some point.

After the advent of writing, lifetimes of experience could be handed down from generation to generation. As long as a civilization is not exterminated, and there exist those who understand its language, this information can persist forever. Such is the case of the Chinese or, with a stretch, the Egyptians. Certainly, it is more difficult to unlock ancient records without living knowledge of the language, but it is not impossible.

Isolated civilizations, whether due to geographic, cultural, or historical reasons, adopted different languages. As civilizations grew, however, and the earth refused to expand, they came into contact, peaceful or otherwise. These interactions spawned a need for communication and with that, translation.

Translation itself is only possible because despite differences among languages, the underlying information is of the same type. Furthermore, language is only a *carrier* of information, not the information itself. With these two observations in mind, you might wonder, if we abstract out "language" and replace it with another medium, such as numbers, can we still encode the same information? Why yes, this is the basis of modern communication. If we are lucky, different civilizations might even communicate the same information in the same language, in which case we have a key to unlock their unknown secrets.

Around the seventh century BC, the Hellenic sphere of influence extended to Egypt, whose culture was gradually impacted by the Greek. After the Greeks (including Macedonians) and Romans became the rulers of Egypt, the Egyptian language was eventually latinized. Hieroglyphics phased out of usage, into the backstage of history. Only temple priests now learned the pictographs, which served only for record keeping. In the fourth century AD, emperor Diocletian eradicated all non-Christianity religions in Egypt, where the knowledge of hieroglyphics ceased to be taught.

Not until 1400 years later, in 1798, did the meaning of hieroglyphics come into light once more. When Napoleon led his expedition to Egypt, he brought along hundreds of scholars. On a lucky day, lieutenant Pierre-Francois Bouchard discovered an ancient Egyptian relic in a placed called Rosetta (Figure 1.3). Atop were inscribed the same message in three languages: ancient Egyptian hieroglyphics, Demotic script (ancient phonetic Egyptian), and ancient Greek. Bouchard immediately realized the importance of his discovery to cracking the hieroglyphic code, so he handed the Rosetta Stone to accompanying scientist Jean-Joseph Marcel. Marcel copied the writings and brought them back to France for further study. In 1801, France was defeated in Egypt, and the physical tablet transferred to British hands, but Marcel's prints were circulated through Europe. Twenty-one years later in 1822, French linguist Jean-Francois Champollion finally decoded the hieroglyphics on the Rosetta Stone. Here, we see that the carrier of those writings, stone or paper, was unimportant. Instead, the writings themselves, the information, were the key.

Rosetta Stone deciphered, the entire history of ancient Egypt, dating back to the 32nd century BC, was suddenly at the disposal of historians and linguists. Thanks to this trove of information, modern day historians know much more about the Egyptians of five thousand years ago than the Mayans of only one thousand years ago. The Egyptians recorded the most important aspects of their lives in writing, so the information lives on. As a natural language processing researcher, I extract two guiding principles from the Rosetta Stone story.

First, redundancy vastly improves the chances that information can be communicated or stored without corruption. The Rosetta Stone repeated the same content three times, so if at least one version remains readable, the stone is decipherable. Fortunately, 2,000 years ago, someone had the foresight to copy Ptolemy's imperial edict in three languages. This concept of redundancy extends to information encoding across noisy channels (for instance, wireless networks). Consider naively that we send the same message twice. Then, it is more likely our recipient will receive at least one of them.

Second, large amounts of bilingual or multilingual language data, known as a corpus, are essential to translation. These data are the bases of machine translation. In this aspect, we do not require more knowledge than Champollion possessed, for the Rosetta Stone. We simply own more computers and apply mathematical tools to speed up the process.

With Rosetta Stone's importance in mind, we should not be surprised that so many translation services and software are all named after Rosetta. These services include Google's own machine translation service, as well as the best-selling (or at least much advertised) software for foreign languages, Rosetta.

We have seen that the emergence of writing was induced by an ancient "information revolution," when people knew more than they could remember. Similarly, the concept of numbers arose when people owned too many possessions to keep track of otherwise. A famous American physicist, George Gamow tells of such a primitive tribe in his book, "One, Two, Three... Infinity." The story goes that two tribal leaders were competing in who could name the largest number. One chief thought for some time and named, "three." After considering for some time, the other chief admitted defeat. Today, a nerdy middle school student might have named a googol, and the second, a googol to the googol, but consider the times. In primitive tribes, all objects were extremely scarce. Beyond three, the tribal chiefs knew only of "many" or "uncountable." By this reasoning, early humans could not have developed a complete counting system.

When our ancestors developed a need for numbers beyond three, when "five" and "eight" became distinguishable, counting systems were invented. Numbers, naturally, are the bases of these counting systems. Like words, numbers were born first in concept, then in writing. With ten convenient fingers to count on, early humans set their numeric systems in base ten. Without a doubt, if we all had twelve fingers instead of ten, we would probably be counting in base twelve right now.

To remember numbers, early humans also carved out scratches on wood, bone, or other portable objects. In the 1970s, archeologists unearthed several baboon leg bones from the Lebombo Mountains, between Swaziland and South Africa. These 42,000-year-old bones featured such scratches, and scientists believe that they are the earliest evidence of counting.

Characters with numeric meaning appeared around the same time as hieroglyphics, thousands of years ago. Nearly all ancient civilizations recorded "one," "two," and "three" in some form of line—horizontal like the Chinese, vertical like the Romans, or wedge-shaped, like the Mesopotamians (Figure 1.4). Early numbers simply recorded information, without any abstract meaning.

Gradually, our ancestors accumulated so much wealth that ten fingers were no longer enough to count their possessions. The easiest method would have been to count on fingers and toes, but then what, grow another set of appendages?

Our ancestors invented far more sophisticated methods, though of course, there may have existed some extinct Eurasian tribe that *did* count on toes.

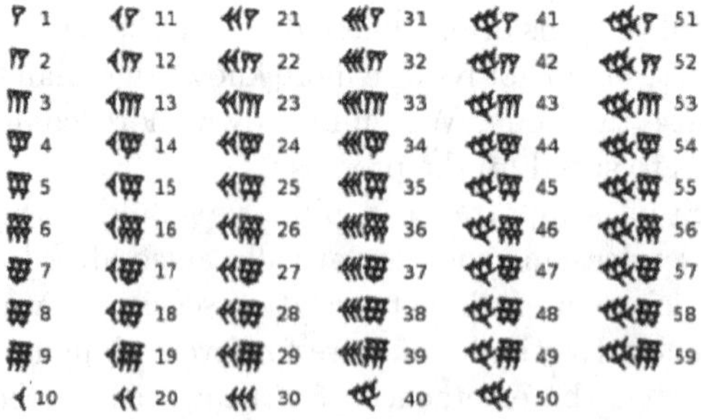

FIGURE 1.4: Cuneiform from ancient Mesopotamia.

They developed the "carry method" in base ten. This system was a great leap forward for mankind: for the first time, man had developed a form of numeric encoding, where different symbols represented different amounts.

Nearly all civilizations adopted the base ten system, but did there exist any civilizations who counted in base twenty—that is, took full advantage of all ten toes before switching to the carry system? The ancient Mayans did so. Their equivalent to our "century" was the sun cycle, 400 years long each. In 2012, the Mayans' last sun cycle ended, and in 2013, the cycle began anew. This I learned from a Mayan culture professor, when I visited Mexico. Somewhere along the way, 2012's "end of the sun cycle" became synonymous with "end of the world." In perspective, imagine if the turn of each century meant apocalypse for us! Of course, we digress.

Compared to the decimal system, a base twenty system comes with nontrivial inconveniences. Even a child, without sophisticated language or vocabulary, can memorize a times tables, up to 9 times 9 equals 81. In base twenty, we would have to memorize a 19 by 19 table, equivalent to a Go board you could say, with 19 times 19 equals 361 entries. Even in burgeoning human civilizations, around 1 AD, no one but scholars would have the mind to study such numbers. The base twenty numeric system, coupled with a painfully difficult writing system, may have significantly slowed the development of Mayan society. Within a single tribe, very few were fully literate.

With respect to the different digits in base ten numbers, the Chinese and the Romans developed distinct units of orders of magnitude. In the Chinese language, there are words for ten, hundred, thousand, 10^4, 10^8, and 10^{12}. In contrast, the Romans denote 1 as I, 5 as V, 10 as X, 50 as L, 100 as C, 500 as D, and 1,000 as M, which is the maximum. These two representations unknowingly captured elements of information encoding. First, different characters represent different numerical concepts; and second, both imply an algorithm for decoding. In ancient China, the rule of decoding was multiplication. Two million is written as two hundred "ten-thousands," or $2 \times 100 \times 10000$. On the other hand, in ancient Rome the rules were addition and subtraction. Smaller numbers to the left meant subtraction, as IV signifies $5 - 1 = 4$. The same numbers to the right meant addition, as VI signifies $5 + 1 = 6$. Unfortunately, the Roman numeric system does not scale well to large numbers. If we wanted to express one million, we would need to write MMMM... and cover up an entire wall (Figure 1.5). Later the Romans invented M with a horizontal bar on top to represent "a thousand times a thousand," but to express one billion, we would still need to cover an entire wall. Therefore, from an efficiency standpoint, the Chinese mathematicians were more clever.

However, the most efficient numeric system came from ancient India, where today's universal "Arabic numerals" actually originated. This system included the concept of "zero," and it was more abstract (hence flexible) than that of both the Romans and the Chinese. As a result, "Arabic numerals" were popularized throughout Europe, which learned of them through Arabic scholars. This system's success lay not only in its simplicity, but also in its

FIGURE 1.5: A Roman mathematician tried to write "one million" on the board.

separation of numbers and words. While a convenience for traders, this detachment led to a divide between natural language and mathematics for thousands of years.

1.3   The mathematics behind language

While language and mathematics grew farther apart as disciplines, their internal similarities did not fade, unaffected by the growingly disparate crowds who studied them. Natural language inevitably follows the principles of information theory.

When mankind established its second civilization in the Fertile Crescent, a new type of cuneiform letter was born. Archeologists first uncovered these symbols on clay tablets, which looked esoterically similar to Egyptian tablets, so they mistook these symbols for pictograms. Soon, however, they realized that these wedge-shaped symbols were actually phonetic, where each symbol stood for a different letter. These constituted the world's earliest phonetic language.* The British Museum owns tens of thousands of these slate and clay tablets, carved with cuneiform letters. These engravings, along with Assyrian reliefs, are among the most valuable Babylonian relics.

---

*If we treat each stroke of a Chinese character as a "letter," we could consider Chinese as "alphabetic" as well, but only in two dimensions.

This alphabetic language was developed by the Phoenicians and introduced to the east coast of Syria, in the west of Mesopotamia. Businessmen and traders, the Phoenicians preferred not to carve intricate wedge letters, so they designed an alphabet of 22 symbols. This alphabet spread with the Phoenicians' business interests, reaching the Aegean islands (including Crete) and the ancient Greeks. Upon reaching the Greeks, the alphabet was transformed into a fully developed alphabet, with no more ties to the Babylonian cuneiform script. Spelling and pronunciation were more closely linked, and the Greek alphabet was easier to learn. In the next few centuries, accompanying Macedonian and Roman conquests, these languages with at most a few dozen characters were embraced by much of Eurasia. Today, we refer to many Western phonetic languages as "Romance languages" for the Romans' role in linguistic dissemination.

Human language took a large leap from hieroglyphics to phonetic languages. An object's description transformed from its outward appearance to an abstraction of its concept, while humans subconsciously encoded these words as combinations of letters. Furthermore, our ancestors chose very reasonable encodings for their languages. For the Romans, common words were often short, and obscure words long. In writing, common words required fewer strokes than uncommon ones. Although our ancestors did not understand information theory, their ideas were fully compliant with the principle of making an encoding as short as possible. The resulting advantage is that writing saves time and material.

Before Cai Lun invented paper, writing was neither easy nor cheap. For example, in the Eastern Han dynasty (around first century AD), text was often inscribed on materials including turtle shell, stone, and bamboo. Since the process was so arduous, every single character was treated as if written with gold. Consequently, classical Chinese writing was painfully concise, while contemporary spoken language was much more verbose and colloquial, not much different from today's Chinese. In fact, the Lingnan Hakka peoples of southern China closely retain the ancient spoken language, with vocabulary and mannerisms of the late Qing dynasty.

This concept of language compression aligns with basic ideas in information theory. When a communication channel is wide, information can be sent directly; but if a communication channel is narrow, then information must be compressed as much as possible before delivery and recovered upon receiving. In ancient times, two people could speak quickly (wide channel), so no compression was needed. On the other hand, writing was slow and expensive (narrow channel), so scholars must first distill daily vernacular into exquisitely crafted poetry. Converting everyday speech into fine writing was a form of compression, and interpreting classic Chinese today is a form of decompression.

We also see this phenomenon in action when streaming video. Broadband provides high bandwidth, so we can watch high-definition videos with sharp resolution. Mobile data plans enforce heavy limits, and data is sent over unreliable networks, so latency is much higher, and resolution is a few magnitudes lower. Though a few thousand years ago there was no information theory, classic Chinese writing adhered to its principles.

Around the time of the late Babylonians, two historical works were produced: one Chinese, one Jewish. Chinese historian Sima Qian wrote a 530,000-word account of Chinese history in the classical style, and the ancient Jews began documenting their history in the Middle East, under Babylonian rule. This latter body of work consisted of Moses' teachings, and we refer to them collectively as the Torah. Its straightforward prose is similar to Sima Qian's writings, but unlike the Chinese manuscript, the Torah was taken into the Bible, whose writing spanned many centuries. Later scribes worked from manuscripts that were hundreds of years old themselves, so copying errors were unavoidable. Scholars say that today, only Oxford University owns an errorless copy of the ancient Bible.

Ancient Jewish scholars copied the Bible with utmost devotion and propriety, washing their hands to pray before writing the words "God" or "Lord." However, copy errors would undeniably emerge, so the scholars devised an error detection method, similar to that used in computers today. They assigned each Hebrew letter to a number, so that every row and column summed to a known value. After coping each page, a scholar would verify that the sums on the new page were the same as those on the old or conclude that he erred in the transcription. Each incorrect sum for a row (or column) signified at least one error on that row, so errors could be easily found and eliminated (see Figure 1.6). Like the ancient Hebrews, modern computers also use the idea of checksums to determine whether data is valid or corrupted.

FIGURE 1.6: Ancient Jewish scholars check every row and sum to verify that they copied the Bible correctly.

From ancient times to now, language has become more accurate and rich, largely due to advances in grammar. I am not a historian of languages, but I would guess that with high probability, grammar started taking shape in ancient Greek. If we consider morphology (constructing words from letters) as the encoding rules for words, then grammar captures the encoding and decoding for languages. However, while we can enumerate all words in a finite collection, the set of possible sentences is infinite. That is, a few tomes worth of dictionary can list all the words in the English language, but no one can compile all English writings ever to exist. Mathematically speaking, while the former can be completely described by a finite set of rules (trivially, we can enumerate all words), the latter cannot.

Every language has its niche usages that grammar rules do not cover, but these exceptions (or "inaccuracies") give language its color. Occasional dogmatic linguists treat these exceptions as "sick sentences." They spend their lives trying to eliminate the linguistic disease and purify the language through new grammar rules, but their work is futile. Take Shakespeare, for instance. Classics now and popular in his time, Shakespeare's works often contained famous, yet ungrammatical phrases. Many attempts were made to correct (or rather, tamper with) his writings, but while these attempts have been long forgotten, Shakespeare's "incorrect" writings persisted. Shakespearean brilliance, taught in schools around the world, originates from grammatical "mistakes."

Grammatical deviancy in literature leads to a controversy: do we consider our existing bodies of text (corpus) as the true expression of language, or should we designate a set of rules as correct usage? After three or four decades of debate, natural language processing scientists converged on the former, that existing data is truth. We will cover this period of history in Chapter 2.

1.4   Summary

In this chapter, we traced the history of words, numbers, and language to pique the reader's appetite for the mathematics intrinsic to our lives. Many of the topics introduced here are the focus of later chapters, including the following.

- Principle of communication and the model of information dissemination

- Encoding, and shortest codes

- Decoding rules and syntax of language

- Clustering

- Checksums (error detection and correction)

- Bilingual texts and corpuses, useful in machine translation

- Ambiguity, and the importance of context in eliminating ambiguity

Modern natural language processing researchers are guided by the same principles as our ancestors who designed language, though the latter's choices were mostly spontaneous and unintentional. Mathematics is the underlying thread through past and present.

Chapter 2

Natural language processing—From rules to statistics

In the previous chapter, we introduced that language emerged as the medium of information exchange between humans. Any language is an encoding of information: it is composed of individual units—letters, words, or Arabic numerals—and an encoding algorithm—a collection of grammar rules. When we communicate an idea, our brains apply grammar encodings on these units to extract comprehensible sentences from abstract thought. If the audience understands the same language, they can use the same rules to decode a string of words back into abstract thought. Thus, human language can be described in terms of information theory and mathematics. While many animals have means of communication, only mankind uses language to encode information.

By 1946, modern computers were becoming increasingly accessible, and computers began to outperform humans at many tasks. However, computers could only understand a limited set of machine-oriented commands, not at all like the English we speak. In this context, a simple question arose: do computers have the potential to understand natural language? Scientists have contemplated this idea since the computer's earliest days, and there are two cognitive aspects to it. First, is a computer powerful enough to understand natural language? Second, does a computer process and learn natural language the same ways a human would? This chapter explores these two questions in depth, but on a high level, the answer to both is a resounding yes.

2.1   Machine intelligence

The earliest proponent of machine intelligence was the father of computer science, Alan Turing. In 1950, he published the seminal paper, "Computing machinery and intelligence," in the journal *Mind*. Rather than detailing new research methods or results, this paper was the first to provide a test that determined whether a machine could be considered "intelligent." Suppose a human and machine were to communicate, both acting as if they were human (Figure 2.1). Then, the machine is intelligent if the human cannot discern whether he is talking to a machine or another human. This procedure is known as the eponymous Turing Test. Though Turing left behind an open question, rather than an

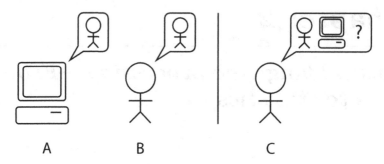

FIGURE 2.1: The human cannot tell whether he is talking to a human or machine, behind that wall.

answer, we generally trace the history of natural language processing back to that era, 60 years ago.

These 60 years of history can be roughly divided into two phases. The first two decades, from the 1950s to the 1970s, were spent in vain but well-intentioned efforts. Scientists around the world wrongly assumed that machines learned the same way humans did, so they spent some nearly fruitless 20 years trying to replicate human thought processes in machines. Only in the 1970s did scientists reassess their approach towards natural language processing, which entered its second phase of research. In the past 40 years, their mathematical models and statistical methods have proven successful. Nowadays, natural language processing is integrated into many consumer products, such as voice assistants on smartphones or automated phone call agents. There are few remaining contributions by the early natural language processing scientists, but their work has been essential to understanding the success of modern methods and avoiding the same pitfalls. In this section, we recount the history of early machine intelligence efforts and in the next, we follow the transition to statistical methods.

Tracing the history of artificial intelligence, we come to the summer of 1956. Four young visionaries—28-year-olds John McCarthy and Marvin Minsky, 37-year-old Nathaniel Rochester, and 40-year-old Claude Shannon—proposed a summer workshop on artificial intelligence at Dartmouth College, where McCarthy was teaching at the time. Joining them were six additional scientists, among which were 40-year-old Herbert Simon and 28-year-old Allen Newell. At these seminars, they discussed unsolved problems in computer science, including artificial intelligence, natural language processing, and neural networks. It was here that the concept of artificial intelligence was first formulated. Other than Shannon, these ten scientists were of little fame or import at that time, but in years to come, four would win Turing Awards (McCarthy, Minsky, Simon, and Newell). Though he received no Turing Award himself, Shannon has an equivalent position to Turing in the history of computer science as the "father of information theory"; in fact, the highest award in information theory is named for Shannon.

These ten scientists were later hailed as the top computer scientists of the 20th century, having initiated a multitude of research areas that are still active today, many of which have directly improved our lives. Unfortunately, the bright minds gathered at Dartmouth that summer produced few results of merit during that month. In fact, their understanding of natural language processing would have been inferior to that of a PhD student today. This is because the scientific community had made a gross misunderstanding about the nature of natural language processing research.

At that time, scientists presumed that in order to complete advanced tasks like translation, a computer must first understand natural language and in turn, possess human-level intelligence. Today, very few scientists insist on this point, but the general public still believes, to some extent, that computers require human-like intelligence to carry out intelligent tasks. In perspective, this assumption is not without reason. For example, we take for granted that a Chinese-English translator is fluent in both tongues. To humans, this is an intuitive deduction. However, this assumption falls into the "flying bird" fallacy. That is, by simply observing birds, we might design an airplane to flap wings in the proper manner, without needing to understand aerodynamics. In reality, the Wright brothers invented the airplane through aerodynamics, not bionics. Exactly replicating mechanisms in the natural world may not be the optimal way to construct competent machines. Understanding natural language processing from a computer's perspective is akin to designing an airplane from aerodynamic principles. For translation, a computer does not need to understand the languages it translates between; it simply translates.

Today, speech recognition and translation are widely adapted technologies, with billions of users, but few understand their underlying mechanisms. Many mistakenly assume that computers understand language, when in reality, these services are all rooted in mathematics, and more specifically, statistics.

In the 1960s, the primary problem scientists encountered was to teach computers to understand natural language. Prevailing beliefs decomposed natural language processing into two tasks: syntactic analysis and semantics extraction. That is, scientists believed that computers could understand natural language by uncovering the grammatical structure of text and looking up the meaning of words. Unfortunately, these goals were misguided, rooted in centuries-old presumptions. Linguistic and language studies have been well established in European universities since the Middle Ages, often forming the core of their curricula. By the 16th century, standardized grammar was becoming widespread, a byproduct of the Bible's introduction outside Europe. By the 18th and 19th centuries, Western linguists had formalized the study of various languages. The large corpus of papers produced by these scholars led to an all-encompassing linguistic system.

In the study of Western languages, we are guaranteed to encounter grammar rules, parts of speech, and word formation patterns (morphology). These rules provide a methodical way to learn foreign languages, and they are straightforward to describe to a computer. As a result, scientists had high hopes for applying traditional linguistics to syntactic analysis.

Compared to syntactic analysis, semantics were much harder to encapsulate and convey to a computer. Until the 1970s, most results about semantic analysis were mediocre at best.* Despite limited successes, semantics are indispensable to our understanding of language, so governments do fund both syntactic and semantic analysis research. The history of bringing natural language processing research to application is described in Figure 2.2.

Let us illustrate an example of syntactic analysis. Consider the simple sentence, "Romeo loves Juliet." We can separate the sentence into three parts: the subject, predicate, and punctuation. Each part can be further incorporated into the following syntactic parse tree.

Computer scientists and linguists often denote these sentence analysis rules as "rewrite rules." The rules used above include:

- sentence → noun phrase + verb phrase + punctuation

- noun phrase → noun

- verb phrase → verb + noun phrase

- noun phrases → nouns

- noun → "Romeo"

- verb → "loves"

- noun → "Juliet"

- punctuation → "."

Before the 1980s, grammar rules for natural language processing were manually created, which is quite different from modern statistical methods. In fact,

Application	Speech recognition	Machine translation	Question-answer	Document summarization
Understanding	Understanding of natural language			
Foundation	Syntactic analysis		Semantic analysis	

FIGURE 2.2: Early attitudes towards natural language processing.

---

*It is worth mentioning that ancient Chinese linguists primarily focused on semantics, rather than grammar. Many ancient monographs, such as "Shuowen Jiezi" (explaining graphs and analyzing characters), were the results of such research.

up until 2000, many companies including the then well-known SysTran still relied primarily on large sets of grammar rules. Today, although Google pursues a statistical approach, there are still vestiges of grammar rules in some products like Google Now (the precursor to Google Assistant).

In the 1960s, compiler technologies were propelled forward by Chomsky's formal language theory. High-level programming languages utilized context-free grammars, which could be compiled in polynomial time (see appendix, Polynomial problem). These high-level programming languages appeared conceptually similar to natural language, so scientists developed some simple natural language parsers in the same spirit. These parsers supported a vocabulary of a few hundred words and allowed simple clauses (sentences with a single digit number of words).

Of course, natural language is far more elaborate than simple sentences, but scientists assumed that computers' explosively increasing computation power would gradually make up for these complexities. On the contrary, more computation power could not solve natural language. As we see in Figure 2.3, syntactic analysis of even a three-word sentence is quite complex. We require a two-dimensional tree structure and eight rules to cover three words. To a computer, these computations are negligibly fast—until we evaluate the growth in complexity with respect to the text. Consider the following sentence from a *Wall Street Journal* excerpt:

"The FED Chairman Ben Bernanke told the media yesterday that $700B bailout funds would be lended to hundreds of banks, insurance companies, and automakers."

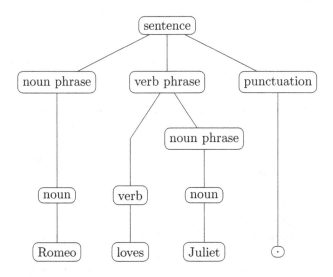

FIGURE 2.3: Syntactic parse tree for "Romeo loves Juliet."

This sentence still follows the "subject, predicate, punctuation" pattern,

> Noun Phrase [The FED Chairman Ben Bernanke] —— Verb Phrase
> [told the media... automakers] —— Punctuation [.]

The noun phrase at top level can be further divided into two noun phrases, "The FED Chairman" and "Ben Barnanke," where the former acts as a modifier to the latter. We can similarly decompose the predicate, as we can any linear statement. That said, the resultant two-dimensional parse tree would become exceedingly complicated, very fast. Furthermore, the eight rules provided for the "Romeo loves Juliet" example cannot adequately parse this new sentence. This complex example not only has more elements to parse, but also requires more rules to cover its structure.

In the general case, there are two barriers to analyzing all natural language using a deterministic set of rules. First, there is an exponential growth in the number of rules (not including part-of-speech tagging rules) required to cover additional grammatical structures. To cover a mere 20% of all statement requires tens of thousands of rules. Linguists simply cannot produce enough rules fast enough, and more specific rules are often contradictory, so context is required to resolve conflicts. If we want to cover over 50% of all statements, then every additional sentence requires many new grammatical rules.

This phenomenon is similar to an adult's reduced capability of picking up foreign languages. Children build language schemas as their brains develop, directly integrating language into the way they think. In contrast, adults learn foreign languages by studying vocabulary and grammar rules. An intelligent adult who learns English as a second language may find it difficult to speak as fluently as a native speaker or perform well on the GRE, even after 10 years of immersion. That is because rules cannot exhaustively describe a language's nuances.

Even if we obtained a full set of rules governing a language, we cannot efficiently implement them as algorithms. Thus, the second barrier to using deterministic rules for natural language processing is computational intractability. Recall that aforementioned compiler technologies of the 1960s parsed context-free grammars, which could be evaluated in polynomial time. These grammars are inherently different from natural language's context-dependent grammar. In natural language, meaning is often derived from surrounding words or phrases. Programming languages are artificially designed to be context-free, so they are much faster to evaluate. Since natural language grammars depend on context, they can become very slow to parse.

The computational complexity (see appendix) required to parse such grammars was formalized by Turing Award winner, Donald Knuth. Context-free grammars could be parsed in $O(n^2)$ time, where n is the length of the statement. On the other hand, context-dependent grammars require at least $O(n^6)$ time. In other words, if we had a sentence of 10 words, then a context-dependent grammar would be ten thousand times slower to parse than context-free grammar.

As sentence lengths grow, the difference in running time explodes. Even today, a very fast computer (Intel i7 quad-core processor) takes a minute or two to analyze a sentence of 20-30 words using rule-based methods. Therefore in the 1970s, IBM, with its latest mainframe technologies, could not analyze useful statements using grammar rules.

2.2   From rules to statistics

Rule-based syntactic analysis (for both grammar and semantics) came to an end in the 1970s. Scientists began to realize that semantic analysis was even more difficult with a rule-based system: context, common sense, and "world knowledge" often contribute to meaning, but are difficult to teach a computer. In 1968, Minsky highlighted the limitations of the then artificial "intelligence" in semantic information processing with a simple example by Bar-Hillel. Consider the two sentences, "the pen is in the box" and "the box is in the pen." The first sentence is easy to understand, and a foreign student who has studied half a year of English can comprehend its meaning. The second sentence may cause confusion for such students—how can a large box fit inside a pen? For a native speaker, the second sentence makes perfect sense alone: the box fits inside a fenced area, a pig pen or similar. Whether "pen" refers to a writing implement or farm structure requires a degree of common sense to determine. Humans develop these sensibilities from real-world experience; unfortunately for scientists, computers do not live" in the real world. This is a simple example, but it clearly illustrates the challenges of analyzing semantics with computers.

Around this time, interest in artificial intelligence waned, partly due to the obstacles inflicted by misguided research. Minsky was no longer an unknown young fellow, but one of the world's leading artificial intelligence experts; his views then significantly impacted the US government's policies on science and technology. The National Science Foundation and other departments were disappointed by the lack of progress in natural language processing, and combined with Minsky's uncertainty, funding in this field greatly shrunk over time. It can be said that until the late 1970s, artificial intelligence research was more or less a failure.

In the 1970s, however, the emergence of statistical linguistics brought new life to natural language processing and led to the remarkable achievements we see today. The key figures in this shift were Frederick Jelinek and his group in IBM's T.J. Watson Research Center. Initially they did not aim to solve natural language, but rather the problem of speech recognition. Using statistical methods, IBM's speech recognition accuracy increased from roughly 70% to about 90%, while the number of words supported increased from hundreds to tens of thousands. These breakthroughs were the first steps towards practical applications of this research, outside the laboratory. We will pick up Jelinek's story later in this chapter.

IBM Watson Laboratories methods and achievements led to a momentous shift in natural language processing. A major figure in these developments was Dr. Alfred Spector, who took roles of VP of research at IBM and Google and held a professorship at Carnegie Mellon University. After Spector joined Google in 2008, I chatted with him about that shift in natural language processing those years ago. He said that while Carnegie Mellon University had delved very deep into traditional artificial intelligence, the researchers there encountered a few insurmountable obstacles. After visiting IBM Watson Laboratories later that year, he realized the power of statistical methods, and even as a professor of the traditional system, he could sense great change on the horizon. Among his students were Kai-Fu Lee, who was among the first to switch from traditional natural language processing methods to statistics. Along this line of research, Kai-Fu Lee's and Hsiao-Wuen Hon's excellent work also helped their thesis advisor, Raj Reddy, win the Turing Award.

As the head of research in two of the world's top technology companies, Spector's keen sense of the future of artificial intelligence was unsurprising. However, this was not a unanimous recognition; the schism between rule-based and statistical-based natural language processing lasted another 15 years, until the early 1990s. During this period, subscribers to each viewpoint organized their own conferences. In mutually attended conferences, each sector held their respective sub-venues. By the 1990s, the number of scientists adhering to the former method steadily decreased, and attendees of their conferences gradually switched over. As such, the prevailing viewpoint slowly converged to statistics.

In perspective, 15 years is a long time for a scientist; anyone who began their doctoral research following traditional methods and insisted on that path may have emerged to realize that their life's work had no more value. So why did this controversy last for 15 years? First, it takes time for a new branch of research to mature. Early on, the core of the statistical approaches lay in communication models and their underlying hidden Markov models (these mathematics are described in more detail later in the book). The input and output of this system were one-dimensional sequences of symbols, whose ordering is preserved. The earliest success of this system was speech recognition, followed by part-of-speech tagging.

Note that this model's output differs significantly from that of traditional methods. Syntactic analysis required a one-dimensional sentence as input but output a two-dimensional analysis tree. Traditional machine translation output a one-dimensional sequence (in another language), but did not preserve semantic order, so the output had little practical value.

In 1988, IBM's Peter Brown proposed a statistical-based machine translation approach, but IBM had neither enough data nor a strong enough model to produce useful results. IBM was unable to account for the variations in sentence structure across languages. For example, in English we say "a red apple," but in Spanish, it would become "una manzana roja," or "an apple red." With these limitations, few studied or made progress in statistical machine translation. Eventually, even Brown joined Renaissance Technologies to make a fortune in investment. Twenty years later, the paper of Brown and his colleagues became popular and highly cited.

On the technical side, syntactic analysis's difficulty lies in the fact that grammatical components are often interspersed within a sentence, rather than simply adjacent to one another. Only a statistical model based on directed graphs can model complex syntaxes, and it was difficult to envision how those would scale. For a long time, traditional artificial intelligence scientists harped on this point: they claimed that statistical methods could only tackle shallow natural language processing problems and would be useless against larger, deeper problems.

In the past three decades, from the late 1980s to the present, the ever-increasing amounts of computing power and data have made the daunting task of statistical natural language processing attainable. By the late 1990s, statistical methods produced syntactic analysis results more convincing than those of the linguists. In 2005, the last bastion of rule-based translation, SysTran, was surpassed by Google's statistics-based methods. Google had achieved a more comprehensive and accurate translation system, all through mathematical models. That is why we can say that mathematics will answer all of natural language processing problems.

Recall that there were two reasons it took traditional natural language processing 15 years to die out. The first was purely technical—models required maturity through time—but the second was practical—scientific progress awaited the retirement of the old linguists. This is a frequent occurrence in the history of science. Qian Zhongshu, in the novel *Besieged City*, remarked that even if scientists in their prime are not physically old, they may hold tightly to old ideas. In this case, we must patiently wait for them to retire and relinquish their seats in the halls of science. After all, not everyone is willing to change their points of view, right or wrong. So the faster these people retire, the faster science can advance. Therefore, I often remind myself to retire before I become too confused and stubborn.

Two groups contributed heavily to the transition between the old and new generations of natural language processing scientists. Other than Jelinek's own IBM-Johns Hopkins collaboration (which included myself), the University of Pennsylvania, led by Mitch Marcus, also played a big role. Marcus managed to obtain support from the National Science Foundation to set up the LCD project, which amassed the world's largest major-language corpus and trained a group of world-class researchers. Scientists from these two groups joined the world's leading research institutions, forming a de facto school of thought and shifting academia's predominant viewpoint.

At the same time, applications of natural language processing have also changed tremendously in the past three decades. For example, the demand for automatic question-answer services has been replaced with web search and data mining. As new applications relied more on data and shallow natural language processing work, the shift towards statistics-based systems was expedited.

Today, there are no remaining defenders of the traditional rule-based approach. At the same time, natural language processing has shifted from simple syntactic analysis and semantic understanding to practical applications, including machine translation, speech recognition, data mining, knowledge acquisition, and so on.

2.3   Summary

With respect to mathematics, natural language processing is equivalent to communication models. Communication models were the missing link between language as an encoding of information and natural language processing, but it took scientists many decades to arrive at this realization.

Chapter 3

Statistical language model

Again and again, we have seen that natural language is a *contextual* encoding for expressing and transmitting information. For computers to understand natural language, mathematical models must first capture context. A model that accomplishes this—also the most commonly used model in natural language processing—is known as the statistical language model. This model is the basis of all natural language processing today, with applications including machine translation, speech recognition, handwriting recognition, autocorrect, and literature query.

3.1 Describing language through mathematics

The statistical language model was created to solve the problem of speech recognition. In speech recognition, a computer must decide whether a sequence of words forms a comprehensible sentence, and if so, return the result to the user.

Let us return to the example from the previous chapter:

> The Fed Chair Ben Bernanke told media yesterday that $700B bailout funds would be lended to hundreds of banks, insurance companies and auto-makers.

This sentence reads smoothly and its meaning is clear. Now suppose we changed the order of some words in the sentence, so that the sentence becomes:

> Ben Bernanke Federal Reserve Chairman of $700 billion told the media yesterday that would be lent to banks, insurance companies, and car companies hundreds of.

The sentence's meaning is no longer clear, but the reader can still infer its meaning, through the numbers and nouns. Now we scramble the words in the sentence to produce:

> the media Ben $700B of told Fed companies that lended yesterday insurance and banks, of auto-makers The and Chair, hundreds would be Bernanke the.

This final sentence is beyond comprehension.

If we ask a layperson to distinguish the differences between the three versions, he might say that the first follows proper grammar and is easy to understand; the second, while not grammatical, still retains meaning through words; the third obfuscates the words and any remaining meaning. Before the last century, scientists would have agreed. They would have tried to determine whether a text was grammatical, and if so, whether it conveyed any meaning. As we discussed in the previous chapter, this methodology led to a dead end. Near the turn of the century, Jelinek changed the prevailing perspective with an elegant statistical model.

Jelinek assumed that a sentence is meaningful if it is likely to appear, where this likeliness is measured with probabilities. We return to the three sentences again. The probability the first sentence appears is about 10^{-20}, while second and third range from 10^{-25} and 10^{-70}, respectively. While these probabilities are all extremely small, the first sentence is actually 100,000 more likely to appear than the second, and billions of billions more likely to appear than the third.

For the mathematically rigorous, let sentence $S = w_1, w_2, \ldots, w_n$ represent an ordered sequence of individual words of length n. We would like to determine the probability $P(S)$ that S appears in any body of text we can find. Naively, we could compute this probability as follows. Enumerate all sentences ever uttered in the entirety of human history, and count the number of times S appears. Unfortunately, even a fool could see the impossibility of such an approach. Since we cannot determine the true probability that any sentence occurs, we require a mathematical model for approximating this value. $S = w_1, w_2, \ldots, w_n$, so we can expand $P(S)$ as

$$P(S) = P(w_1, w_2, \ldots, w_n). \tag{3.1}$$

By conditional probability, it follows that the probability w_i occurs in the sequence is equal to the probability w_i appears alone, multiplied by the probability that w_i appears, given the existing sequence $w_1, w_2, \ldots, w_{i-1}$. We thus expand $P(S)$ to

$$\begin{aligned} P(S) &= P(w_1, w_2, \ldots, w_n) \\ &= P(w_1) \cdot P(w_2|w_1) \cdot P(w_3|w_1, w_2) \ldots P(w_n|w_1, w_2, \ldots, w_{n-1}), \end{aligned} \tag{3.2}$$

where each word's appearance depends on the previous words.*

From a calculation standpoint, $P(w_1)$ is easy to find, and $P(w_2|w_1)$ is not hard either. However, $P(w_3|w_1, w_2)$ provides some difficulty because it involves three variables (words), w_1, w_2, and w_3. To calculate each variable's probability, we require a table the size of a language dictionary, and the size of these tables

---

*More accurately, we would say $P(w_1)$ is the probability w_1 appears at the *start* of a sentence.

for conditional probabilities increases exponentially with the number of variables involved. When we reach the final word w_n, we realize that $P(w_n|w_1, w_2, \ldots, w_{n-1})$ is computationally infeasible to calculate. So we return to our mathematical model—is there any way to approximate this probability without these expensive computations?

Russian mathematician Andrey Markov (1856-1922) proposed a lazy but effective method for resolving this quandary. Whenever we want to determine the probability of word w_i in sentence S, we no longer consider all previous words $w_1, w_2, \ldots, w_{i-1}$. Rather, we reduce our model to only consider the previous word w_{i-1}. Today we call these models Markov chains.* Now, the probability that sentence S appears becomes

$$P(S) = P(w_1) \cdot P(w_2|w_1) \cdot P(w_3|w_2) \ldots P(w_n|w_{n-1}). \tag{3.3}$$

Equation 3.3 corresponds to the statistical language model known as the bigram model. Of course, we may also consider the $n-1$ words preceding word w_n, and such grams are known as N-gram models, introduced in the next section.

After applying the Markov assumption to $P(S)$, we encounter the next problem of determining conditional probabilities. By definition,

$$P(w_i|w_{i-1}) = \frac{P(w_{i-1}, w_i)}{P(w_{i-1})}. \tag{3.4}$$

Nowadays, researchers can easily calculate the joint probability $P(w_{i-1}, w_i)$ and the marginal probability $P(w_i)$. Extensive amounts of digital text (corpuses) allow us to count occurrences of word pairs (w_{i-1}, w_i) and individual words w_i. Suppose we have a corpus of size N words. Let $\#(w_{i-1},w_i)$ be the number of times w_{i-1}, w_i appear consecutively, and $\#(w_{i-1})$ be the number of times w_{i-1} appears alone. We now obtain relative frequencies of occurrences in our corpus.

$$f(w_{i-1}, w_i) = \frac{\#(w_{i-1},w_i)}{N} \tag{3.5}$$

$$f(w_{i-1}) = \frac{\#(w_{i-1})}{N} \tag{3.6}$$

By the Law of Large Numbers, as long as our sample size N is large enough, then relative frequencies of events approach their true probabilities.

$$P(w_{i-1}, w_i) \approx \frac{\#(w_{i-1},w_i)}{N} \tag{3.7}$$

$$P(w_{i-1}) \approx \frac{\#(w_{i-1})}{N} \tag{3.8}$$

---

*Markov first proposed this assumption in 1906, but Andrey Kolmogorov extended this idea to infinite state spaces in 1936.

Returning to conditional probability, we notice that the two above values share the denominator N, so $P(w_i|w_{i-1})$ simplifies to a ratio of counts.

$$P(w_i|w_{i-1}) = \frac{\#(w_{i-1}, w_i)}{\#(w_{i-1})} \tag{3.9}$$

Now, some readers may begin to appreciate the beauty of mathematics, which converts complicated problems into elementary forms. This process may seem slightly unbelievable at first. Basic probabilities can accomplish feats including speech recognition and machine translation, in which complicated grammars and artificial intelligence rules stumble. If you are questioning the efficacy of the model we just presented, you are not alone. Many linguists also doubted this statistics-based approach. In the modern day, however, researchers have found that statistics trumps the most sophisticated rule-based systems. We provide three real-world applications of this model, for the disbelievers.

Thirty years ago, the ex-president of Google China, Kai-Fu Lee, meteorically rose to the preeminence of speech recognition. As a doctoral student, he built a continuous-speech, speaker-independent system using the same Markov models described above.

Now let us fast forward some years to the 21st century, when Google developed its machine translation system, Rosetta. Compared to many universities and research institutions, Google was late to the machine translation game. Prior to Rosetta, IBM, the University of Southern California, Johns Hopkins, and SysTran had already made much progress in the field. For many years, these veteran institutions had participated in the National Institute of Standards and Technology (NIST) machine translation evaluation, a widely recognized performance benchmark. Google's Rosetta system first entered the competition in 2005, after only two years of development. To everyone's astonishment, Rosetta came in first, by a wide margin. It surpassed all rule-based systems, upon which over a decade of research was focused. What's Google's secret weapon? It is datasets and mathematical models hundreds of times larger than those of its competitors.*

A few years after that, around 2012, we come to some of my own work at Google involving the automatic question-answer system. At the time, many scientists in laboratory settings could design computers that answered basic questions. That is, "what is the population of the United States," or "what year was Donald Trump born?" While factual questions had straightforward answers like 300 million or 1946, "why" and "how" questions required further explanation. Questions like "why is the sky blue" were beyond the scope of computers then. We could mine large datasets to compile an answer with all the right keywords, but in order to sound fluent, a computer required the statistical language model. Since incorporating that model, this system has been released in English and other foreign languages. If you Google "why is the sky blue" today, you will

---

*We refer to the number of n-tuples available.

find a well-prepared answer. In future chapters, we will delve into the details of this system's mathematics, but now we return to the statistical language model.

From these three examples, we see that the statistical language model has become an indispensable part of any "intelligent" computer system. Nonetheless, there are still countless implementation details between the mathematics and the software. For example, if a word (or pair of words) we encounter does not appear in the corpus, or only appears rarely, then our estimated probabilities are skewed. Fortunately, Jelinek and his colleagues not only presented this model, but also filled in many of its corner cases. These cases will be outlined in the following section. If you do not work with the statistical language model or find this mathematics unpalatable, then worry not. We have already laid all the foundations for the statistical language model, and you do not need to read further to appreciate its brilliance. The beauty of mathematics lies in its potential to accomplish groundbreaking work with a simple model.

3.2 Extended reading: Implementation caveats

Suggested background knowledge: probability theory and statistics.

Most of this book's chapters will include an extended reading section, tailored towards professionals and those who wish to further study behind the mathematics behind the presented topics. These sections may require additional background to fully comprehend, so to save the reader's time, I have noted the suggested prerequisite knowledge at the start of each section. While not a hard requirement, familiarity with these subjects will improve the reading experience. Later chapters do not depend on these extended readings, so readers are free to choose whether to read or skip any such section.

3.2.1 Higher order language models

In Equation 3.3 of the previous section, we assumed that the probability of each word w_i is related to its immediately preceding word w_{i-1}, but unrelated to all other preceding words w_j, where $j < i - 1$. The reader might wonder whether such an assumption is slightly oversimplified. Indeed, it is easy to find such examples where word w_i depends on words other than w_{i-1}. Consider the phrase, "sweet blue lilies." Here, "sweet" and "blue" both describe "lilies," but only "blue" is considered the previous word. As a result, we should perhaps consider the previous two words. Generalizing even further, we may modify our assumption so that a word w_i depends on several preceding words.

We express this assumption in mathematical notation. Suppose word w_i depends on $N-1$ preceding words. We modify the conditional probability of w_i, given an existing sequence of words, to

$$P(w_i|w_1, w_2, \ldots, w_{i-1}) = P(w_i|w_{i-N+1}, w_{i-N+2}, \ldots, w_{i-1}). \quad (3.10)$$

Equation 3.10 corresponds to a higher order Markov chain, or in natural language processing terms, an n-gram model. Some special cases of the n-gram model include $n = 2$, the bigram model (Equation 3.3), and $n = 1$, the context-free unigram model. The unigram model assumes that each word's appearance is unrelated to the appearance of nearby words. In practice, the most commonly used model is $N = 3$, the trigram model, and higher orders beyond $N = 3$ are rarely used.

After learning that context is key to language, we might ask, why limit the orders to such low orders? There are two reasons. First, N-gram models quickly become computationally intractable. Both the spatial and time complexity of a meta model undergo exponential growth as the number of dimensions increases. A language's words are usually contained in a dictionary V where $|V| = 10,000$ to 100,000 words. Each additional dimension adds exponentially many word combinations, where order N would require $O(|V|^N)$ space and $O(|V|^{N-1})$ time. The model significantly improves from $N = 1$ to $N = 2$ and slightly improves at $N = 3$, but the model experiences little additional benefits when increasing to $N = 4$. While the marginal benefit of increasing N approaches none, the resources consumed increases exponentially, so $N = 4$ is near the maximum order that anyone will use. Google's Rosetta translation and voice search systems use the 4-gram model, but the model requires over 500 servers to store.

Second, we cannot cover all language phenomena, even if we try to increase N to infinity. In natural language, the relevance of context may span paragraphs, or even sections. Literary analysis is a prime suspect of far-ranging context, given that symbols may appear throughout an entire novel. Even if we *could* increase the order of our model to $n = 100$, we could not encapsulate all context. Such is a limitation of the Markov assumption. In its place, we have tools that consider long distance dependencies, which we will discuss later in this book.

3.2.2   Training methods, zero-probability problems, and smoothing

As described in Equation 3.3, the statistical language model requires the knowledge of all conditional probabilities, which we denote as the model's parameters. Obtaining these parameters from a corpus's statistics is known as training. For example, the bigram model requires two numbers, $\#(w_{i-1}, w_i)$ and $\#(w_{i-1})$, to estimate the conditional probability $P(w_i|w_{i-1})$. Seemingly straightforward, the model fails if the pair (w_{i-1}, w_i) never appears. Does $P(w_i|w_{i-1})$ equal 0 for all new sentences? Likewise, if the two counts are the same, does $P(w_i|w_{i-1})$ really equal 1? These questions involve the reliability of our statistics.

So far, we have assumed that statistics observed on our sample (available corpus) are equivalent to those that could be observed on the population (entire language). As long as our datasets are large enough, the Law of Large Numbers guarantees that this assumption holds. For example, suppose we want to determine the demographics of customers at the local mall. On a Saturday afternoon, we count 550 women and 520 men, so we conclude that $550/(550 + 520) = 51.4\%$ are women, and 48.6% are men. However, suppose we are in a hurry, so we walk in and out of the mall on a Tuesday morning. There are only 5 people, 4 men and 1 woman. Is it reasonable to conclude that 20% of customers are women, and 80% are men? What if we only observe 3 people, all women? Do we dare conclude that not a single man shops at the mall? Of course not. Small samples produce statistics with high variance.

Many people would say the unreliability of small samples is common sense, but the same people often forget this fact when training language models. Instead of addressing the lack of data, they question the model's validity. Today, the statistical language model has withstood the test of time, and many digital communications applications are built atop similar models. Given that the theory is sound, the remaining challenge lies in proper training of the model.

The first solution that comes to mind is to directly increase our dataset size. With some quick estimations though, we will see that all the data in the world is not enough to compose an adequate sample. Take Chinese, for instance. A Chinese dictionary contains around 200,000 words, not all commonly used.* To train a trigram model, we require $200,000^3 = 8 \cdot 10^{15}$ unique parameters. If we obtain our corpus by crawling the Chinese web, there are around 10 billion web pages, each with an average of 1,000 words (overestimate). So even if we manage to obtain all of these pages, we still have only 10^{13} parameters. Therefore, if we use direct ratios to calculate conditional probabilities (Equation 3.3), most of these probabilities will still be zero. Such a model is deemed "not smooth." Even though most Chinese words rarely appear in practice, we cannot pretend that the zero-probability issue does not exist. As a result, we must find an alternative way to address the issue of sparse data.

Scientists would love to possess the complete boxed set of language data, but such data is unattainable. Thus, the art of training useful models arises from identifying the optimal workarounds for limited sample sizes. Under the guidance of Alan Turing, British mathematician I.J. Good (1916-2009) developed a probabilistic method for weighing different data points and accounting for missing data. From the outset, we assert that the data sample is incomplete, so we would lower the weights of "untrustworthy" data and give those weights to "unseen events." This method was later dubbed the Good-Turing frequency estimate, for approximating unknown probabilities.

Although its calculations are slightly cumbersome, the Good-Turing estimate is easy to understand in principle. We allocate a small portion of the total probability mass to events that are possible in theory, but were not observed in the data sample. Probabilities of all events must add to 1, so the probabilities of

---

*According to Google IME.

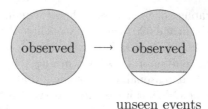

unseen events

FIGURE 3.1: Left: all observed events. Right: a portion of the probability mass is given to unseen events.

observed events are decreased, with the most unlikely events decreased the most (see Figure 3.1).

As an example to the reader, we will walk through the Good-Turing calculations with a sample statistical dictionary. Suppose there are N_r words that appear r times in a corpus of size N. Then it is evident that

$$N = \sum_{r=1}^{\infty} rN_r. \tag{3.11}$$

Without considering words that were not observed, we would estimate the relative frequency of words appearing r times as $f_r = rN_r/N$. If r is small, then the statistics for those words are unreliable. Instead of f_r, the Good-Turing estimation uses a smaller factor d_r as the probability estimate, where

$$d_r = \frac{(r+1)N_{r+1}}{N_r} \tag{3.12}$$

and

$$\sum_r d_r \cdot N_r = N. \tag{3.13}$$

Generally speaking, there are more words that appear once than twice, and more words that appear twice than three times. This inverse relationship between r and N_r is known as Zipf's law. Figure 3.2 illustrates this trend for a small sample corpus.

We see from Figure 3.2 that N_r decreases as r increases, or equivalently, $N_{r+1} < N_r$. Due to this inverse relationship, the adjusted probability d_r is generally less than f_r, and since $d_0 > 0$, the unseen event is given a very small, non-zero probability. In practice, we also set an additional threshold a for f_r, above which we do not lower the probability to d_r. With this adjustment, words that occur with frequencies above a retain their original probability estimates. For those below a, unlikeliness is exaggerated to make room for unseen values. As a result, low probabilities are penalized, zero probabilities are raised to small positive

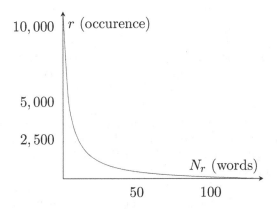

FIGURE 3.2: Zipf's law. Relationship between r (occurrences in corpus) and N_r (number of such words).

values, and the total probability mass does not change. The final probability distribution is smoother than the original.

Smoothing is easily extensible to word pairs. Given bigram pair (w_{i-1}, w_i), we would like to estimate $P(w_i|w_{i-1})$. The initial ratio estimates give that

$$\sum_{w_i \in V} P(w_i|w_{i-1}) = 1. \qquad (3.14)$$

Applying the Good-Turing Discount to pairs, we would reduce the weights of unlikely pairs (w_{i-1}, w_i) such that $\sum_{w_i \in V} P(w_i|w_{i-1}) < 1$. This reduction leaves more probability mass for unseen pairs. Now we can express the probability adjustments as follows,

$$P(w_i|w_{i-1}) = \begin{cases} f(w_i|w_{i-1}) & \text{if } \#(w_{i-1}, w_i) \geq T \\ f_{\text{G-T}}(w_i|w_{i-1}) & \text{if } 0 < \#(w_{i-1}, w_i) < T \\ Q(w_{i-1}) \cdot f(w_i) & \text{otherwise} \end{cases} \qquad (3.15)$$

where T is a threshold count usually around 8-10, $f_{\text{G-T}}$ is the Good-Turing estimate, and $Q(w_{i-1})$ is the adjusted zero-probability value,

$$Q(w_{i-1}) = \frac{1 - \sum_{w_i \text{ seen}} P(w_i|w_{i-1})}{\sum_{w_i \text{ unseen}} f(w_i)}. \qquad (3.16)$$

These three cases ensure that Equation 3.14 holds. The above regularization method was first proposed by IBM scientist S.M. Katz, and it is known as the Katz backoff model. This estimation scheme is also extensible to trigram models as follows.

$$P(w_i|w_{i-2}, w_{i-1}) = \begin{cases} f(w_i|w_{i-2}, w_{i-1}) & \text{if } \#(w_{i-2},w_{i-1},w_i) \geq T \\ f_{G-T}(w_i|w_{i-2}, w_{i-1}) & \text{if } 0 < \#(w_{i-2},w_{i-1},w_i) < T \\ Q(w_{i-2}, w_{i-1}) \cdot f(w_i|w_{i-1}) & \text{otherwise} \end{cases}$$

$$(3.17)$$

To address 4-gram and higher-order models, we add more variables to the conditional probabilities. Herman Ney and his colleagues developed methods to optimize the Katz backoff method, but the fundamentals are the same. We direct interested readers towards reference 2.

Recall the inverse relationship between a word's occurrences in text r and the number of such words N_r. Single words appear more than pairs, so by the Law of Large Numbers, the relative frequency ratios for single words are closer to the true probability distribution. Likewise, estimates for pairs are more accurate than for triples, and so on. At the same time, the zero-probability problem is less severe for lower order models than higher order models. On the other hand, higher order models provide more contextual information. In the past, scientists leveraged this property as an alternative method of smoothing; linear interpolation between low order and high order models was known as deleted interpolation. Equation 3.18 demonstrates this method, where the λ values are positive and sum to 1. Compared to the Katz backoff model, deleted interpolation performs slightly worse, so it is not widely used today.

$$P(w_i|w_{i-2}, w_{i-1}) = \lambda(w_{i-2}, w_{i-1})f(w_i|w_{i-2}, w_{i-1})$$
$$+ \lambda(w_{i-1})f(w_i|w_{i-1}) + \lambda f(w_i)$$

$$(3.18)$$

3.2.3   Corpus selection

Besides training methods, another crucial aspect of statistical models is the training data, or the selection of a corpus. If the training corpus and the actual application data are drawn from different sources, then the model's performance will be reduced.

Imagine that we are building natural language processing software for web search. Then we should use web source code and search queries as training data, rather than an archive of *The New York Times*, even if the web data are messy and digital newspapers are pristine. Common sense, you might say, but China company Tencent once made the mistake of using China's equivalent of *The New York Times*, the state newspaper. Data from web crawling are notoriously messy, ill-formatted, and unstandardized. Digital archives, on the other hand, are carefully prepared and curated. Tencent thought that newspaper data were cleaner and thus, better for training. With the newspaper implementation, mismatches arose between search queries and web content. After switching to web data instead, Tencent's search engine vastly improved, despite the low quality of the latter data.

Besides heeding the data source, we generally prefer to gather as much data as possible. Although smoothing alleviates some of the low or zero-probability issues, model parameters are still estimated better with more data. Higher order models possess more parameters to be trained and require more data to train them. In an ideal world, we would have unlimited data, not all applications have access to much data at all. Take machine translation, for instance. There exist very few bilingual texts, and even fewer digital ones. Here, a one-sided pursuit of a high order model is pointless and other methods may be required.

When we have obtained enough training data consistent with our application, we begin to consider the effects of noise on our model. Often, data must be preprocessed and massaged into a more cohesive form. Data contains two types of noise, random and systematic. Random noise (human error) is customarily ignored, as removal costs are high and benefits are low. Systematic noise, which may obscure the data patterns in question, are easier to handle and largely dealt with. For instance, HTML tags in a web page should be removed so that only content remains. Therefore, noise removal should be performed when costs are low compared to benefits. Future chapters will deal further with the importance of training data in models.

3.3   Summary

Tucked away inside many of the world's "intelligent" technologies, the statistical language model reduces hard language problems to basic mathematical representations. Though simple and easy to understand, this model captures deep mathematical insights. An expert can spend many years studying the intricacies of this model, of which we provided a brief taste in the extended reading. Alternatively, any layperson can appreciate this model's elegance, knowing that it powers many of our life conveniences today; herein lies the charm and appeal of mathematics.

Bibliography

1. Turing, Alan (October 1950), "Computing Machinery and Intelligence", Mind LIX (236): 433–460, doi:10.1093/mind/LIX.236.433.

2. Katz, S. M. (1987). Estimation of probabilities from sparse data for the language model component of a speech recogniser. IEEE Transactions on Acoustics, Speech, and Signal Processing, 35(3), 400–401.

3. Jelinek, Frederick, Statistical Methods for Speech Recognition (Language, Speech, and Communication), 1998 MIT Press.

4. Kneser, R. and Ney, H. 1995. Improved backing-off for m-gram language modeling. In Proceedings of ICASSP-95, vol. 1, 18–184.

Chapter 4

Word segmentation

I had initially titled this chapter "Word segmentation in Chinese," since the original word segmentation problem was aimed towards Chinese. As a generic problem, however, word segmentation has applications in many Asian languages (e.g., Japanese, Korean, Thai), as well as in syntax analysis of phonetic languages (e.g., English, French). Thus, the Chinese word segmentation we introduce here is only an example to illustrate the overarching problem.

4.1 Evolution of Chinese word segmentation

The statistical language model was built atop texts as collections of words, because words are the smallest units of semantics below which meaning is obscured. Words in Western phonetic languages are clearly delimited, often by spaces, so statistical models are directly applicable to these languages. In contrast, many Asian languages have no clear delimiter between its words.* For these languages, researchers need to extract words from a soup of characters before further processing.

An input to word segmentation takes the form of a string of consecutive characters without delimiters. English is written with spaces between words, but we can remove them to emulate Chinese. Suppose we input

DonaldTrumpcamefromafamilyofbusinessmen.

Its corresponding output is a sequence of words, delimited by a character we specify. With a slash as the delimiter,

Donald Trump / came / from / a / family / of / businessmen.

If asked how to produce the latter from the former, you would probably propose the earliest method invented: a dictionary scan. This method was first proposed by Professor Nanyuan Liang of the Beijing University of Aeronautics and Astronautics. As expected, we traverse a sentence from left to right, adding

---

*Chinese and Japanese have no spaces between words in sentences. In Korean, there is a delimiter between noun phrases and verbs, but no delimiters within noun phrases themselves.

characters to a word until it matches some word in the dictionary. When we encounter compound words like "Donald Trump," we add as many words to the noun as possible. Though seemingly naive, this method can correctly delimit the above example. We first encounter "Donald," which is a valid name on its own, but then we see "Trump," and we combine the two into "Donald Trump." Adding the character "c" renders the word invalid, so we know that "c" belongs to the next word "came," and so on.

Keep in mind that the thousands of Chinese characters are all valid words on their own, unlike in our alphabet, so the dictionary scan method can actually delimit 70-80% of all Chinese sentences. Given its low cost, this method actually performs pretty well! In the 1980s, Dr. Xiaolong Wang of the Harbin Institute of Technology further improved the dictionary scan method with the least word segmentation theory. Still, given any complicated sentence, this approach will inevitably stumble. For example, suppose we wanted to share a restaurant location with the phrase:

PaloAltoUniversityAvenue.

Matching words left to right for the longest word, we would read out,

Palo Alto University / Avenue,

rather than the correct,

Palo Alto / University Avenue.

Furthermore, this method does not solve for double-meanings. What if we *did* mean a building named "Avenue" inside "Palo Alto University"? We require more sophisticated tools to address these challenges.

We introduced in the first chapter that ambiguity has developed alongside language since the dawn of human civilization. In ancient China, the expenses of writing condensed classical thought into texts only readable by dedicated scholars. Across the Western world, contrasting interpretations of the same religious texts spawned a multitude of religious denominations. Ambiguity is inherent to language.

Before the 1990s, many scholars at home and abroad attempted to solve word segmentation with the equivalent of grammar rules. As the title of the second chapter may suggest, these scholars were met with little success. Some scholars began to discover the power of statistical models, but none correctly applied the theory to this problem. Finally around 1990, Dr. Jin Guo at Tsinghua University's Department of Electrical Engineering successfully addressed the ambiguity problem via the statistical language model. He reduced the Chinese word segmentation error rates by an order of magnitude, and the above examples can all be solved with his model.

Jin Guo was the first person from mainland China to achieve success by deliberately using the statistical language model. Alongside his tireless work, his background in information theory played a significant role in his discoveries. Although Guo held a doctorate degree in computer science, many of his long-time research colleagues studied digital communications. We can consider Guo as a Chinese counterpart to Jelinek, and it is unsurprising that Guo also applied communication models to natural language processing.

We can summarize statistical word segmentation through a bit of mathematical notation. Suppose each sentence S can be segmented a number of ways. Let us take three segmentations for simplicity,

$$(A_1, A_2, \ldots, A_k)$$
$$(B_1, B_2, \ldots, B_m)$$
$$(C_1, C_2, \ldots, C_n)$$

where each variable $A_1, A_2, \ldots, B_1, B_2, \ldots, C_1, C_2, \ldots$, etc. is an individual word, and k, m, n may be distinct numbers. That is, a sentence may be segmented into different meanings with different numbers of words. The best segmentation is also the most probable, so

$$P(A_1, A_2, \ldots, A_k) > P(B_1, B_2, \ldots, B_m)$$

and

$$P(A_1, A_2, \ldots, A_k) > P(C_1, C_2, \ldots, C_n).$$

Building on top of the previous chapter, as long as we can determine the probabilities of each sentence, we can select the best segmentation.

As before, the devil lies in the implementation. If we enumerated all potential segmentations and calculated the probability of each, our computer would probably crash from exhaustion. Instead, we treat sentence segmentation as a dynamic programming problem and apply the Viterbi algorithm to quickly find the best choices (algorithm provided in later chapters). On a high level, we have illustrated the process below in Figure 4.1.

After Guo's work, many scientists used the statistical language model to improve word segmentation. Worth mentioning are Professor Maosong Sun of

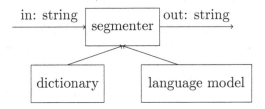

FIGURE 4.1: Statistical language model for word segmentation.

Tsinghua University and Professor Dekai Wu of Hong Kong University of Science and Technology (see reference 4). Professor Sun solved the word segmentation problem without a dictionary, while Professor Wu applied Chinese word segmentation to English phrases. Wu then linked corresponding Chinese and English phrases for machine translation, so we see that word segmentation has applications beyond Chinese itself.

On a side note, linguists regard words with more room for contention. For example, some people consider "University of California Berkeley" a single word, while others might assert that "Berkeley" describes "University of California." After all, college students casually say "UC Berkeley," but official documents write "University of California, Berkeley" with a comma. As a compromise, a computer may treat such cases with a nested structure. Algorithms first recognize the larger word group "University of California Berkeley," then break down the phrase into "University of California" and "Berkeley." This method was first published in *Computational Linguistics*, and many systems have adopted the practice since.

Typically, different applications often require varying granularities of word segmentation. Machine translation may prefer a coarser granularity, so we treat "University of California Berkeley" as a single phrase. On the other hand, speech recognition requires smaller units, since humans speak word by word. In its early years, Google did not have enough employees to develop a word segmentation system of its own, so it used a third party system by Basis Technology. As a result, the results were not optimized for search. Later, with the resources to grow, Google hired Dr. Xianping Ge and Dr. An Zhu to build a word segmentation system tailored specifically towards search. Results improved, and the search team was happy.

At this point, many readers may be thinking, Western languages do not have a word segmentation issue! The spacebar is the longest key on our keyboards for a reason. Unfortunately, fancy new technologies pose new problems. Handwriting recognition introduces a sort of segmentation problem: people do not leave consistent spaces between their letters, so computers end up reading the equivalent of "DonaldTrumpcomesfromafamilyofbusinessmen." Thus, word segmentation algorithms, developed for Chinese, are applicable to English too.

In fact, many of the mathematical algorithms in natural language processing are independent of language. At Google, when we designed a new algorithm for a certain language, we strongly weighed the algorithm's extensibility to a variety of other languages. Only then could we support hundreds of languages in a global search product.

Robust as statistical models are, any method has its limitations. Though word segmentation is carried out with high accuracy, we can never achieve perfection because language is inherently—and intentionally—ambiguous. Statistical language models are built from existing data, which is largely in accordance with the opinion of the masses and the most common usages of language. In special cases, public opinion may be "wrong." For instance, poetry derives depth from layered meanings, and a computer cannot explicate a poem the same way a writer will. Fortunately, such cases are rare in everyday text, so they cause us no concern.

I conclude this section with two points on word segmentation. First, word segmentation is considered a solved problem in natural language processing. In industry, the basic statistical language model and several well-known techniques combine to produce quality results without much additional effort. There is limited room for improvement, so segmentation is not an area of active research. Second, Western languages typically lack the need for word segmentation, but the increasing popularity of tablet PCs and smartphones has increased the usage of handwriting input. As humans, we do not prefer to write with large, clear spaces for a computer's ease, so word segmentation is beneficial to implement for handwriting applications. While it originated from a need to process Chinese texts, word segmentation is useful to many languages. As is the theme of this book, data-backed mathematics provided an easy answer to this problem, and now we will know what to blame if our tablets do not recognize our handwriting.

4.2 Extended reading: Evaluating results

4.2.1 Consistency

Measuring the quality of word segmentation results is harder than it seems. Often we would like to compare a computer's results with a human's opinion and see if they align. In this case, however, even humans do not agree on right and wrong. Some people may say that "Tsinghua University" is a compound word, while others would argue that "Tsinghua" and "University" should be divided. Either arrangement comes with its justifications, and linguists have their rules for determining which is better. Different applications may even call for one usage or the other.

A distinction between compound words may seem small, but the disparity in human judgment is nontrivial. In 1994, I worked with IBM to recruit 30 Tsinghua sophomores for a Chinese word segmentation study. For consistency, we trained the students for half an hour before the study, and we provided 100 representative Chinese sentences for them to dissect. Despite the students' respectable education and training, they only agreed 85-90% of the time. To provide context, the world's best handwritten-digit classifiers consistently perform error rates under 1%. Therefore, word segmentation has a low consistency rate among humans.

Human discrepancies are unavoidable in the gold standard for this task, but we can still distinguish between good and bad systems. Before the statistical language model was applied, the gap between human and computer was relatively large, and computers clearly performed more poorly. When the differences between human and computer became comparable to that between any two humans, computers reached a limit on further improvement.

With the statistical language model, the variation among word segmentation algorithms was much less than among between humans. Consequently, it became meaningless to compare computers with humans. After all, we cannot presume that a 97% accurate model is superior to a 95% accurate one; the accuracies depend on quality of humans annotating the data. We can only gather that one model is more in line with its human data than the other. Seeing that improvement space is minimal, researchers consider word segmentation a solved problem. As long as it utilizes the statistical language model, any computer can segment words with high performance.

4.2.2   Granularity

Humans will disagree about word segmentation because they implicitly think with different word granularities. In Chinese and many other languages, the smallest unit of expression is the word. Just as the atom is to matter, the word is to language: if we split an atom, its chemical properties will no longer hold, and if we split a word, it becomes meaningless. To the point, linguists are in agreement. For example, no one claims that "California" should be split into "Cal / i / for / nia." On the contrary, "University of California Berkeley" elicits conflicting viewpoints. Some people view this as a noun phrase, and others, a single proper noun. The contention here arises from disagreement about the fundamental "particle size" for words.

Instead of crowning a victor of correct granularity, we realize that each application has its own optimal granularity. In machine translation, for instance, large granularities generally improve results. The Chinese translation for "Lenovo" involves four characters, two of which mean "association" on their own. Translating Lenovo into "association company" makes little sense when we know the four characters mean "Lenovo" on the whole. Other applications, such as web search, tend towards smaller granularities. Searches for "University of California Berkeley" should include results that only mention "Berkeley," which might exist in the context of "UC Berkeley." If we exclude these pages, then the search engine misses a lot of valid results.

To support multiple granularities, we could build several distinct word segmentation systems, but this approach is wasteful and unnecessary. Rather, it is more efficient for a single system to support a multitude of granularity levels. In other words, we would specify whether we want "University of California Berkeley" or separate phrases. Granularity selection is typically straightforward to implement, as described below.

We first require a dictionary of basic words ("California," "Berkeley") and a dictionary of compound words ("University of California Berkeley"). Next, we train languages models L_1, L_2 using these two dictionaries, respectively.

Using the basic dictionary, we analyze the sentence with L_1 to obtain the fine-grained segmentation. By the conventions in Figure 4.1, the input is a sentence and the output is a word sequence. Basic words are unequivocally words, so other than updating the dictionary now and then, we do not require additional research.

Afterwards, we use the compound dictionary and apply L_2 to generate coarse-grained segmentation. Figure 4.1's input is a word sequence from L_1, and the output is a sequence with compound words or phrases. Note that the only change between L_1 and L_2 segmentation is the dictionary. None of the algorithms change between different granularities; we simply swap out dictionaries.

Understanding granularity allows us to evaluate the accuracy of word segmentation systems in more depth. Segmentation inconsistencies can be split into two groups: errors and granularity incongruities. Errors can be further classified as boundary errors ("University of / California Berkeley") and word errors ("Cal / i / for / nia"). Obvious to human readers, errors should be eliminated to improve accuracy. Granularity differences are less severe and can be attributed to human judgment. In fact, most of the disagreement between human segmentation falls into the granularity category. Usually, we do not regard these disagreements as severe segmentation errors, so we will not train our segmentator to specifically reduce those errors. For some applications, it is more important to determine as many valid segmentations as possible, rather than selecting a single correct answer.

In the long term, the only way to maintain useful word segmentation systems is to constantly mine new data and refresh our dictionaries and phrase lists. Indeed, word segmentation work in recent years mostly comprises of staying relevant.

4.3   Summary

After decades of research, Chinese word segmentation is now a solved problem with the statistical language model. Though human error will always exist, we have learned how to distinguish the good segmentation systems from the bad. Quality depends both on data and implementation.

Bibliography

1. Nanyuan Liang, Automatic word segmentation of Written Chinese, http://www.touchwrite.com/demo/LiangNanyuan-JCIP-1987.pdf.

2. Jin Guo, Statistical Language Models and Chinese word/spelling conversion studies. http://www.touchwrite.com/demo/Guojin-JCIP-1993.pdf.

3. Jin Guo. 1997. Critical tokenization and its properties. *Comput. Linguist.* 23, 4 (December 1997), 569–596.

4. Sun Maosong, Shen Dayang, and Benjamin K. Tsou. 1998. Chinese word segmentation without using lexicon and hand-crafted training data. In *Proceedings of the 17th international conference on Computational linguistics - Volume 2* (COLING '98), Vol. 2. Association for Computational Linguistics, Stroudsburg, PA, USA, 1265–1271. DOI: https://doi.org/10.3115/980432.980775.

5. Dekai Wu. Stochastic inversion transduction grammars, with application to segmentation, bracketing, and alignment of parallel corpora. IJCAI-95: 14th Intl. Joint Conf. on Artificial Intelligence, 1328–1335. Montreal: Aug 1995.

Chapter 5

Hidden Markov model

The hidden Markov model (HMM) is a marvelous mathematical tool. Theoretically uncomplicated, this model works wonders in practice; it is widely regarded as the most efficient solution to most natural language processing problems, including speech recognition and machine translation. After we illustrate how the HMM tackles these various applications, I am sure you will take to this model with affection as well.

5.1 Communication models

As we introduced in the first two chapters, humans have continually built means of information exchange through the dawns and dusks of civilization. Language is but such a tool, so there are unbreakable ties between natural language and information theory. In fact, the essence of communication is just the coding and transmission of information. When early linguistic efforts focused on grammar rules and semantics, the distance increased between language studies and information theory. Upon returning to communication models in the 20th century, researchers finally began to solve many age-old problems about natural language.

Let us paint a high-level picture of the typical communication system. A sender transmits a signal through some medium to a receiver. For example, a cellphone can send voice signals over the air to a cell tower, or various data through a wire to a computer. In a broad sense, the sender first encodes some information and then transmits that over a channel. Afterwards, the receiver decodes the received data into the original information, thereby completing the information exchange. Visually, Figure 5.1 depicts the six elements of communication, as defined by Roman Jakobson (1896-1982).*

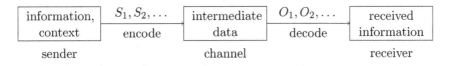

FIGURE 5.1: Communication model.

---

*These elements are the sender, channel, receiver, message, context, and coding.

In the above figure, the sequence $S = S_1, S_2, \ldots$ represent the sent signal (e.g., broadcasted by cellphone), and $O = O_1, O_2, \ldots$ represent the received signal (e.g., received by cell tower), or the observations. After decoding, the receiver can convert O back into S, or the original information.

At this point, we will take a step back from the specifics of communication models and connect our understanding of natural language processing. Communication models provide an alternative angle to natural language processing. For example, speech recognition is merely a case of speculating on what a speaker says. Here, the receiver analyzes an audio signal and extracts the ideas a person conveys. The sender, then, is that person's brain, which communicates through the medium of vocal cords and air. If the receiver is a computer, then we dub this process "automatic speech recognition." If instead, the receiver is another human's ear, we gloss over this exchange as everyday talking.

Similarly, many natural language processing applications fall under the communication model abstraction. In Chinese-English translation, the sender speaks in Chinese over an English-encoded channel, and the receiver's task is to decipher the underlying information. Likewise, if the sender transmits garbled-up English with many spelling mistakes and the receiver distills clear English from the mess, we term the process autocorrect. If we think hard enough, we realize that most natural language problems can be cast to a communication model.

Now that we have seen the universality of communication models, we fill in another puzzle piece: how do we efficiently decode O into the original information S? We return to mathematics. The most *probable* decoding is probably correct.

We can express this idea with conditional probabilities. We denote the sequence S that maximizes $P(S_1, S_2, \ldots | O_1, O_2, \ldots)$ as

$$S = \arg \max_{\text{all } S} P(S_1, S_2, \ldots | O_1, O_2, \ldots) \tag{5.1}$$

where arg is short for argument, or the value of S that attains the maximum probability. Of course, calculating this probability is not trivial either, but we can find the value indirectly. Using Bayes' formula, we can convert Equation 5.1 into

$$P(S|O) = \frac{P(O_1, O_2, \ldots | S_1, S_2, \ldots) \cdot P(S_1, S_2, \ldots)}{P(O_1, O_2, \ldots)}. \tag{5.2}$$

where $P(O|S)$ is the probability that sent message S will become O upon reception, $P(S)$ is the probability a sender sends S, and $P(O)$ is the probability of receiving O.

At this point, you might wonder if we have actually *increased* the complexity of the problem, as the equations are growing longer. Do not worry, for that is not the case. We will show that these new quantities are actually easier to find.

First, the received message $O_1, O_2, \ldots$ stays the same throughout our analysis, so its probability $P(O_1, O_2, \ldots)$ is a constant. Since we are solving a maximization problem, we can discard constants, and we are left with

$$P(O_1, O_2, \ldots | S_1, S_2, \ldots) \cdot P(S_1, S_2, \ldots). \tag{5.3}$$

There are still two remaining quantities, but we can determine both through hidden Markov models.

5.2   Hidden Markov model

The hidden Markov model was not actually invented by 19th century Russian mathematician, Andrey Markov, but by American mathematician, Leonard E. Baum, and colleagues in the 20th century. Established through a series of papers, the hidden Markov model's training methods are also named for Baum (Baum-Welch algorithm).

To understand the hidden Markov model, which is not a misnomer, we must first introduce the Markov chain. By the 19th century, probability theory had evolved from static to dynamic—from the study of individual random variables to that of stochastic processes (i.e., a sequence of random variables $S_1, S_2, S_3, \ldots$). Philosophically, human understanding took an enormous leap. Mathematically, there were several details to hammer out.

We make the following observations. First, at any time t, the state of the system S_t is random. We illustrate this idea with a hypothetical (and fantastical) example. Suppose you are going on vacation to Neverland, so you would like to know the highest temperature every day. Tinkerbell, the local meteorologist, assures you the weather will be warm and temperate. Unfortunately, the sky is raining buckets when you arrive, and you realize that the weather is unpredictable and random.

Second, the value of the state S_t may be related to any number of nearby states. Unbeknownst to you, the clueless traveler, Neverland had been cold and cloudy for days prior to your arrival. Tinkerbell had predicted high temperatures, but had you known the previous states, you might have packed more warmly.

Now in all seriousness, there are two degrees of uncertainty to this random process. Since any uncertainty is difficult to formalize, Markov proposed a simplification: the distribution of each state S_t depends only on the previous state S_{t-1}. That is, $P(S_t | S_1, S_2, \ldots, S_{t-1}) = P(S_t | S_{t-1})$. In the case of Neverland, we would have Tinkerbell predict each day's weather depending only on the previous day's. As with all models, this assumption may not be appropriate for all problems, but it does provide an approximate solution to a previously difficult problem. Today, we call this simplification the Markov assumption, and stochastic processes consistent with this hypothesis are named Markov

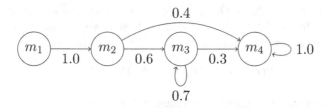

FIGURE 5.2: Markov chain.

processes, or Markov chains. The following figure graphically depicts a discrete Markov chain.

In this Markov chain, the four circles represent four stages, and the edges each represent a possible state transition (see Figure 5.2). For example, there is only one edge between m_1 and m_2 with a weight of 1.0, so if we are ever at state m_1, then we can only move to state m_2 with probability 1.0. Now suppose we reach m_2. We have two options, m_3 and m_4, with transition edge weights 0.6 and 0.4, respectively. These weights correspond to the transition probabilities,

$$P(S_{t+1} = m_3 | S_t = m_2) = 0.6$$
$$P(S_{t+1} = m_4 | S_t = m_2) = 0.4.$$

We can think of the Markov chain as a machine, which randomly selects a starting state S_0 and begins transitioning according to the above rules. After running for T time steps, we receive a sequence of states $S_1, S_2, \ldots, S_T$. With this sequence, we can easily count the occurrences $\#(m_i)$ of each state m_i and the number of each transition $\#(m_i, m_j)$. Then we can predict the transition probability from m_i to m_j as

$$P(S_{t+1} = m_j | S_t = m_i) = \frac{\#(m_i, m_j)}{\#(m_i,)}.$$

Each state S_{t+1} is only related to the previous state S_t, regardless of how we transitioned there. In the above Markov chain, we will always have a 0.3 chance of moving to m_4 from m_3, whether we just came from m_2 or have been circling back to m_3 for several iterations.

The hidden Markov model is an extension of the above Markov chain with an additional caveat: at any time t, we cannot view the state S_t. As a result, we cannot infer transition probabilities or other parameters through the sequence of states $S_1, S_2, \ldots, S_t$. However, at each time t, the hidden Markov model exports a symbol O_t, an observation that depends only on S_t (Figure 5.3). This property is known as the output independence assumption. Though there may exist an underlying Markov chain $S_1, S_2, \ldots$, we see only the outputs. Thus, Baum aptly named this model a "hidden" Markov model.

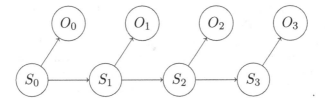

FIGURE 5.3: Hidden Markov model.

Based on the Markov assumption and the output independence assumption, we can calculate the probability that a given state sequence $S_1, S_2, \ldots, S_T$ will produce outputs $O_1, O_2, \ldots, O_T$.*

$$P(S_1, S_2, \ldots, O_1, O_2, \ldots) = \prod_t P(S_t|S_{t-1}) \cdot P(O_t|S_t) \qquad (5.4)$$

You may notice that Equation 5.4 looks a lot like Equation 5.3. This similarity is not a coincidence. We can extend Equation 5.3 to a stochastic process with the two hidden Markov model assumptions, so that

$$P(O_1, O_2, \ldots | S_1, S_2, \ldots) = \prod_t P(O_t|S_t)$$

$$P(S_1, S_2, \ldots) = \prod_t P(S_t|S_{t-1}). \qquad (5.5)$$

In this form, Equation 5.3 is massaged into a hidden Markov model. Furthermore, notice that $P(S_1, S_2, \ldots)$ can be determined by the statistical language model, the focus of our previous chapter. Many other natural language processing problems can be expressed in terms of these same building blocks, so we can apply hidden Markov models to them all. With a tractable expression to optimize, we can use the Viterbi algorithm (details in later chapters) to determine the most probable message.

Various problems in natural language processing all share the same basis, albeit with different names. Speech recognition calls its approach the "acoustic model," machine translation claims its "translation model," and autocorrect utilizes the "correction model." Underneath, these models are all hidden Markov models.

Historically, researchers first attained success with the hidden Markov model in the 1970s, for speech recognition. Under Jelinek, IBM Watson's James and Janet Baker[†] applied this model to speech recognition. Compared to classical artificial intelligence, the Bakers' implementation reduced the speech recognition

---

*I try to avoid excessive mathematical notation, but for concision, $\prod_i f(x)$ is equivalent to the product of $f(x)$ evaluated at all values of i. Product notation is the multiplication equivalent of summation notation, $\Sigma_i f(x)$.

[†]They founded the company Dragon, now obsolete.

error rate by two-thirds, from 30% to 10%. Another development took place in the 1980s, when Kai-Fu Lee created the world's first large-vocabulary, continuous speech recognition system, Sphinx, at Carnegie Mellon University. Later, the hidden Markov model was applied to machine translation, and from there, to autocorrect, handwriting recognition, image analysis, genomics, and an explosion of IT fields. In the past 20 years, even Wall Street has adopted similar models.

I first learned of the hidden Markov model in an ill-reputed class at Tsinghua University, Random Processes. At the time, I did not think twice about the model—I found it cute and basic, but not terribly useful. Later, however, my research advisor gave me a mountain of natural language processing papers to read, and two authors stuck out: Jelinek and Lee. I was amazed at how they applied hidden Markov models to their respective fields, and it was then that mathematics first seduced me with its elegant charm.

5.3   Extended reading: HMM training

Suggested background knowledge: probability theory.

There are three problems to consider when training a hidden Markov model:

1. Given a model, determine the probability of any received message O.

2. Given a model, determine the most probable sent message S that produced received message O.

3. Given enough data, estimate the parameters of the hidden Markov model.

The solution to the first problem is the Forward-Backward algorithm, which this book does not cover. I direct interested readers to Frederick Jelinek's *Statistical Methods for Speech Recognition (Language, Speech, and Communication)*.* As mentioned before, the second problem's solution is the Viterbi algorithm, discussed later in the book. Finally, in this section we will explicate the third problem.

To train a hidden Markov model, we need to estimate two properties. First, transition probability $P(S_t|S_{t-1})$ describes the probability that given state S_{t-1}, we will move to S_t at the next time step. Second, generation probability $P(O_t|S_t)$ gives the probability that at state S_t, the underlying Markov chain will output symbol O_t. Both are known as parameters of the hidden Markov model, and we refer to training as the process of honing in on them.

---

*The MIT Press (January 16, 1998).

Let us begin analyzing these two parameters with condition probabilities. From Bayes' formula, we know the following.

$$P(O_t|S_t) = \frac{P(O_t, S_t)}{P(S_t)} \tag{5.6}$$

$$P(S_t|S_{t-1}) = \frac{P(S_{t-1}, S_t)}{P(S_{t-1})} \tag{5.7}$$

We examine each quantity in turn. We can easily approximate Equation 5.6 if we have enough human annotated data. Put simply, we divide the number of times state S_t outputs O_t by the number of times S_t appears, such that

$$P(O_t|S_t) \approx \frac{\#(O_t, S_t)}{\#(S_t)}. \tag{5.8}$$

Since we rely on manually annotated data to train our model, this method is known as supervised learning. Similarly, the transition probability in Equation 5.7 can be estimated using annotated data counts, except its condition probability is equivalent to that of the statistical language model. Thus, we can use the same formula,

$$P(w_t|w_{t-1}) \approx \frac{\#(w_{t-1}, w_t)}{\#(w_{t-1})}. \tag{5.9}$$

As you may have noticed, we have based these estimates on the assumption that there is a large corpus of human annotated data, ready for use. Unfortunately, many applications, including speech recognition, lack such data. People are unable to determine the state of an acoustic model, so they cannot annotate training data. For other applications, data annotation is feasible but cost-prohibitive. In machine translation, bilingual corpuses are required, and sentences in the two languages must be aligned by bilingual speakers, one by one, before they can be used to train translation models. Obviously, this cost is unbearable. We would require massive control corpuses in two languages, and translation is both expensive and slow. Instead, we prefer to train a hidden Markov model to deduce parameters from observable signals. Without annotated training data, this method (Baum-Welch algorithm) is a form of unsupervised learning.

When estimating state from outputs, we run into a slight hurdle: if two distinct Markov chains can produce the same string of outputs, then which chain is better? Moreover, what if many such suitable Markov chains exist? In these situations, we return to probability. Given distinct models M_{θ_1} and M_{θ_2}, where θ are the parameters, there always exists M_{θ_2} that is more likely to produce output O than M_{θ_1}. Here, the Baum-Welch algorithm steps in to determine the most likely model $M_{\hat{\theta}}$. Its general gist is as follows.

Initially, we start with any model M_{θ_0} capable of producing output O. Note that there must exist many such models; if we set all transition and output probabilities to the same value, we can produce any output string. Then, we continually find better models. Assuming that we have an initial model, we can not only determine the probability $P(O|M_{\theta_0})$, but also run through the Markov chain many times to find all potential pathways to O. In the process, we acquire large amounts of transition and output data, which we can treat as an "annotated" dataset. Now, we apply Equations 5.6 and 5.7 to calculate a new set of parameters θ_1, thus concluding a single iteration. It can be proven that we strictly improve over time,

$$P(O|M_{\theta_1}) > P(O|M_{\theta_0}). \tag{5.10}$$

Afterwards, we start from M_{θ_1}, find a better model M_{θ_2}, and continue until our model no longer improves significantly. These are the fundamentals of the Baum-Welch algorithm. For the exact algorithm, I direct interested readers towards reference 2.

Each iteration of the Baum-Welch algorithm continually updates our model parameters so that the probability of our outputs (an objective function) is maximized. In general terms, this algorithm is an expectation-maximization (EM) process. EM processes are guaranteed to converge to a local optimum, though not the global optimum. Consequently, in natural language processing applications such as paired annotation, supervised learning models often perform better than the unsupervised Baum-Welch algorithm. However, if the objective function is convex, as in information entropy, there is only one global optimum, and the EM process will converge to this value. In Chapter 26, "God's algorithm: The expectation maximization algorithm," we will discuss EM processes in greater detail.

5.4   Summary

Though the hidden Markov model was originally designed for communications, it has since been extended to speech and language processing, as a bridge between natural language processing and information theory. At the same time, the hidden Markov model is an extremely useful machine learning tool. Like many such models, the hidden Markov model requires a training algorithm (e.g., Baum-Welch algorithm) and a decoding algorithm (e.g., Viterbi algorithm). If we can master these two algorithms for our desired applications, then we can add the hidden Markov model to our everyday toolbox.

Bibliography

1. Baum, L. E.; Petrie, T. (1966). "Statistical Inference for Probabilistic Functions of Finite State Markov Chains". The Annals of Mathematical Statistics 37 (6): 1554–1563.

2. Baum, L. E.; Eagon, J. A. (1967). "An inequality with applications to statistical estimation for probabilistic functions of Markov processes and to a model for ecology". Bulletin of the American Mathematical Society 73 (3): 360–363.

3. Baum, L. E.; Sell, G. R. (1968). "Growth transformations for functions on manifolds". Pacific Journal of Mathematics 27 (2): 211–227.

4. Baum, L. E.; Petrie, T.; Soules, G.; Weiss, N. (1970). "A Maximization Technique Occurring in the Statistical Analysis of Probabilistic Functions of Markov Chains". The Annals of Mathematical Statistics 41: 164–171.

5. Jelinek, F.; Bahl, L.; Mercer, R. (1975). "Design of a linguistic statistical decoder for the recognition of continuous speech". IEEE Transactions on Information Theory 21 (3): 250.

Chapter 6

Quantifying information

Thus far, we have said much about information, but information itself remains an abstract concept. We can assert that a text conveys a lot or a little information, but in the end, we still are not sure how much. For example, how much information does a 500,000-character Chinese book contain, and what about a complete set of Shakespeare's plays? Likewise, we often claim that information is useful, but in what way, and to what quantifiable extent? For thousands of years, no one had a satisfying theoretical answer to these questions.

Not until 1948 did Claude Shannon finally present us with an answer. In his famous book, *A Mathematic Theory of Communication*, Shannon proposed the idea of "information entropy," which has since allowed us to measure the amount of information and quantify its value.

6.1 Information entropy

The amount of information a message contains is directly related to its uncertainty. If we want to understand an unpredictable statistic, or if we know nothing about the topic at hand, then we require lots of information. On the other hand, if we already know a lot, then any additional message brings minimal improvements. From this perspective, we claim that the amount of information is equal to the current degree of uncertainty.

Now, we devise a method to quantify uncertainty. Suppose that we avidly desire the results of the 2018 World Cup, but we do not have access to the proper television channels. Instead, a devious man approaches us, offering to answer any yes-no question for $1 each. With little cash in our pockets, we must find the cheapest way to determine the victor. We can maximize information gained for our bucks in a divide and conquer fashion. First, we number every country from 1-32. Then, we ask, "does the winning team lie in teams 1-16?" If so, we narrow the range to 1-8. If not, we know that the champion is one of 17-32, so we ask about teams 17-24. In this way, we will know the answer in five questions, so the World Cup championship question is only worth $5.

Instead of dollars, Shannon used the unit of a bit. In computers, a bit is a binary number, and a byte is 8 bits. In the World Cup question, the amount of information provided by knowing the champion is 5 bits. If instead, we allowed 64 teams to complete, then the answer would be worth 6 bits.

By now, you may have noticed that the number of bits is equal to the logarithm of the number of possibilities ($log\ 32 = 5$ and $log\ 64 = 6$).*

For those familiar with soccer, you may suggest that we do not necessarily require all 5 guesses to determine the victor. After all, some teams like Spain, Brazil, or Germany are much more likely to win, compared to teams like Japan, South Africa, or Iceland. Thus, we do not need to divide all 32 teams into random, equal groups. We could instead separate those likely to win from everyone else. In this way, we might only require 3 or 4 guesses, if we are lucky. Therefore, when teams have different chances of winning, the amount of information encapsulated by the World Cup winner is actually less than 5 bits. In fact, Shannon would state that the exact amount of information is

$$H = -(p_1 log\ p_1 + p_2 log\ p_2 + \cdots + p_{32} log\ 32) \tag{6.1}$$

where p_i represents the probability that team i will win. Shannon dubbed this measurement "entropy," which is usually denoted with an H with units of bits. If teams are equally likely to win (or we know nothing), then the corresponding information entropy is 5. Interested readers are encouraged to derive this value. With a basic mathematical background, you can also prove that the above quantity cannot exceed 5.

For an arbitrary random variable X (e.g., the championship team), its corresponding information entropy is

$$H(X) = -\sum_{x \in X} P(x) log\ P(x). \tag{6.2}$$

Less certain variables may have smaller $P(x)$ values over more possible x, so more information is required. Now as a side note, why is this measure termed "entropy"? Information entropy is quite similar to thermodynamic entropy, which we will discuss in this chapter's extended reading.

Now that we understand entropy, we can return to our earlier question: how much information does a 500,000-character book contain? Well, there are approximately 7,000 commonly used Chinese characters, and if they equally appear, then each character would occupy around 13 bits. However, the most common 10% of characters account for over 95% of all appearances. Even if we do not consider the occurrences of characters in the context (e.g., some characters commonly appear after one another) and only consider individual characters, the entropy of each is around 8-9 bits. Adding in context, we find that each character will only take up around 5 bits. Half-a-million characters will use around 2.5 million bits. With a decent compression algorithm, we can squish the book down to a 320KB file. If we directly use the two-byte Unicode, the file will take around 1MB, or three times the compressed size. The gap between these two values is known as "redundancy" in information theory.

---

*Unless otherwise stated, all logarithms in this book are base 2.

We should note that 2.5 million bits is an average value, and the actual information weighs heavily on its size. A book that repeats a lot of its content will naturally take up less space.

Languages differ in their degree of redundancy, and compared to other languages, Chinese has relatively low redundancy. Often, a book that is translated from English to Chinese becomes a lot thinner in Chinese, or a lot thicker, vice versa. This is consistent with many linguists' view that Chinese is the most succinct language. Those interested in Chinese language entropy can read the article "Complexity of Chinese information entropy and language model" that I coauthored with Professor Zuoying Wang in the *Journal of Electronics*.*

6.2 Role of information

Since ancient times, information and the elimination of uncertainty have been inextricably linked. In some situations, single bits of information propelled powerful consequences. Let us travel back to World War II Soviet Union. When the Nazis lay siege to Moscow, Stalin had few troops to spare in Europe. He could not commit the 600,000 troops stationed in Siberia, because he was unsure whether Japan would attack north, to Russia, or south, to the United States. If Japan chose south, then Stalin could withdraw troops from Siberia to reinforce Moscow. In reality, Japan *did* choose south, as reflected at Pearl Harbor, but Stalin dared not act without this knowledge, for dire repercussions of guessing wrong. Finally, the famed spy, Richard Sorge, sent back a single bit of priceless information: south. The former Soviet Union transferred all Siberian troops to the European front, and the rest is history.

If this story's principle of information can be abstracted and universalized, it can be summarized in Figure 6.1.

Any event (such as the WWII story above) has an inherent degree of randomness, or uncertainty *U*. When we receive information *I* from the outside

information

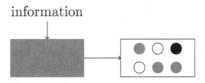

FIGURE 6.1: Information is the only way to eliminate uncertainty from a system. In the absence of any information, a system is a black box. When we introduce information, we learn more and more about the system's inner workings.

---

*http://engine.cqvip.com/content/citation.dll?id=2155540.

(such as the Japanese cabinet decision), we become more certain about the event. If $I > U$, then uncertainty is eliminated; otherwise, we have a new level of uncertainty,

$$U' = U - I. \tag{6.3}$$

On the flip side, if we have no information, then neither complicated algorithm nor formula can reduce uncertainty. This basic fact is often overlooked. Sometimes, "experts" will draw conclusions from non-existent information. It is best to be wary of reports from such "experts." Regardless, nearly all applications of natural language understand, information processing and signal processing are related to the elimination of uncertainty.

Early readers of this book suggested that they wanted to hear more about mathematics in web search. Our next example is precisely that. Web search involves sifting through billions of web pages to find the few most relevant to a user's search query. Billions of pages involve a lot of uncertainty, while the few final results involve little to no uncertainty. Therefore, web search is also a process of using information to eliminate uncertainty.

When the information provided is vague, uncertainty still looms. For instance, common search terms like "United States" or "economy" will match a wider range of pages than a user cares to know about. The user is unlikely to be satisfied, because he probably had a specific aspect in mind. At this point, the correct approach is to tap the remaining resources about the pages themselves, such as information quality. If we still require more information, the user may have to choose which pages are useful to himself. These are the bases of web search.

Among search engines, a common pitfall is to draw circles and fancy transformations around the search keywords. We are not introducing new information by changing the form of search keywords, so this practice has little effect. In fact, this approach is a major reason that some people who labor over their search engines come to no avail.

Even worse a practice is to introduce artificial assumptions about search results. This will render a search engine no different from a random page selector. If we listen to the opinions of some users, then search results will be tailored towards their tastes, and the majority of users will experience a massive decrease in search relevance. In context of our WWII story, this would be as if Stalin chewed over the idea of Japanese invasion at breakfast, then decided upon a military strategy based on the taste of his coffee. Instead of imposing human judgment, the key to search is strategic use of information.

The more information we know about a random event, the less its uncertainty. This information can be directly related to the event, such as the Japanese cabinet decision, or it can be peripheral, like web pages' quality. Peripheral information tells us about other random events, which in turn, relate back to our main question. For example, the statistical language models of previous chapters exhibit both types of information. The unigram model tells us information about the words themselves, while the bigram and higher-order models use contextual information to learn more about the words' usages in a sentence. Using mathematics, we can rigorously prove that relevant information eliminates entropy, but to that end, we must first introduce conditional entropy.

Let X and Y be two random variables, where X represents the event we care about. Given the probability distribution $P(X)$, we can find the entropy of X,

$$H(X) = - \sum_{x \in X} P(x) \log P(x). \tag{6.4}$$

Now suppose we also know the joint probability $P(X, Y)$ that the events occur together, and the conditional probability $P(X|Y)$. We define conditional entropy as

$$H(X|Y) = - \sum_{x \in X, y \in Y} P(x, y) \log P(x|y). \tag{6.5}$$

In the extended reading, we will prove that $H(X) \geq H(X|Y)$, or that after observing Y, the uncertainty about X decreases. With regards to the statistical language model, we can regard Y as the preceding word to X. Then, this inequality implies that the bigram model has less uncertainty than the unigram model. Similarly, we can extend conditional entropy to three variables,

$$H(X|Y, Z) = - \sum_{x \in X, y \in Y, z \in Z} P(x, y, z) \log P(x|y, z), \tag{6.6}$$

where $H(X|Y) \geq H(X|Y, Z)$. That is to say, trigram models have even less uncertainty than bigram models, and so on.

Finally, we come to the interesting question of equality. In the equations above, the potential for equality suggests that we can introduce more information without reducing uncertainty. Though not immediately apparent, this situation is certainly possible. If we receive completely irrelevant information, then we learn nothing more about our topic at hand. For example, if Stalin's spies reported about German and British skirmishes in North Africa, then no matter how detailed the intelligence, the information does not reduce Moscow's dilemma.

To summarize this section in a sentence: information exists to eliminate uncertainty, and natural language processing seeks to extract the correct information for each problem.

6.3   Mutual information

When information and events are related, we eliminate uncertainty—but how do we define this unscientific notion of "related"? Common sense tells us that some random events and random variables are related, while others are not. Knowing that it rained today in Boston would give us a clue about Boston's humidity for the past 24 hours. Rain in San Francisco tells us little about Boston's humidity, but we cannot eliminate the possibility that they are somehow, tenuously linked.* To this point, Shannon proposed the idea of "mutual information," a quantitative measure of relevance between two random events.

Given two random variables, X and Y, their mutual information is

$$I(X; Y) = \sum_{x \in X, y \in Y} P(x, y) log \frac{P(x, y)}{P(x)P(y)}. \tag{6.7}$$

If you are intimidated by long mathematical equations, don't be. All you need to remember is that $I(X; Y)$ stands for the mutual information of X and Y.

Given some random event X, we quantify its uncertainty as its entropy, $H(X)$. Now, suppose we already know the outcome of another random event Y. We can calculate the conditional entropy $H(X|Y)$, or the uncertainty in X, having observed Y. Then mutual information is the difference

$$I(X; Y) = H(X) - H(X|Y). \tag{6.8}$$

So the relevance of two events X and Y can be measured by the reduction in uncertainty of X, after knowing Y. Upon quick inspection, we see that mutual information is bounded between 0 and the minimum of their entropies. When two events are completely related, $I(X; Y)$ is equal to $H(X)$. On the other hand, if X and Y have nothing to do with each other, their mutual information is 0.

In natural language processing, it is easy to compute the mutual information of linguistic features. As long as we have enough data, we can easily estimate $P(X)$, $P(Y)$, and $P(X|Y)$. We use these values to calculate $I(X; Y)$ by Equation 6.7. In the natural language context, mutual information is simply the correlation of speech phenomena.

One of the largest challenges in machine translation is ambiguity—that a single word can take on many meanings. For example, the word "Bush" can be taken as low-lying shrubbery, or alternatively, the former President of the United States. Similarly, the word "Kerry" can be translated as a rare breed of Irish dairy cow, or as Bush's contender in 2004. See Figure 6.2.

What is the right way to disambiguate presidents from plants, then? An intuitive but ineffective solution is grammar rules. Grammatically, both Bush and bush are nouns, whether the writer intends them to be last names or shrubs. We could, of course, add a rule that "presidential names must be parsed as people first," but that is a slippery slope. After adding in all the world

---

*Though by the butterfly effect, the influences may be larger than we imagine.

FIGURE 6.2: Bush and Kerry presidential debates. The "shrubbery" president and the "dairy cow" senator.

leaders, past and present, we should also consider that countries can be leaders themselves, as in the United Nations' members. Then there are various leadership titles, including Secretary General, Prime Minister, Queen, Duke...the combinations are endless.

The simpler solution is mutual information. First, we determine words that have high mutual information with President Bush. Those may include "Washington," "United States," or "President." We repeat the process for the plant type, bush. Similar words may contain "soil," "environment," or "wildlife." With these two sets of words, we look at the context around the word "bush" and decide which "bush" the text is talking about. This method was first proposed by William Gale, Kenneth Church, and David Yarowsky in the early 1990s.

As a doctoral student at the University of Pennsylvania under Mitch Marcus, Yarowsky spent much time collaborating with Bell Labs and other notable researchers. Perhaps eager to graduate, Yarowsky came up with the above method for disambiguation. Simple and clever, this method was astonishingly effective. Its success granted Yarowsky his Ph.D. after only three years, when on average, his peers spent six.

6.4 Extended reading: Relative entropy

Suggested background knowledge: probability theory.

Information entropy and mutual information form a basis for information theory, which is vastly important for natural language processing.

In this section, we introduce another important concept, relative entropy (also known as Kullback-Leibler divergence), and its uses in natural language processing.

Similar to mutual information, relative entropy also measures correlation. While mutual information quantifies the relationship between two random variables, however, relative entropy measures the similarity between two positive-valued functions. It is defined as

$$KL\left(f(x)\|g(x)\right) = \sum_{x \in X} f(x) log\, \frac{f(x)}{g(x)}. \tag{6.9}$$

Again, you need not memorize this formula, but the following properties are useful to know:

1. If two functions are identical, their relative entropy is 0.

2. For two functions, the larger their relative entropy, the more different they are. Likewise, the smaller their relative entropy, the more similar they are.

3. For probability distributions or probability density functions, if their target values are positive, then their relative entropy measures the difference between the distributions.

Note that relative entropy is *not* symmetric. That is,

$$KL\left(f(x)\|g(x)\right) \neq KL\left(g(x)\|f(x)\right).$$

As such, this asymmetry is quite inconvenient. To leverage the properties of relative entropy and maintain symmetry, Jensen and Shannon proposed that we use the average of the two sides to obtain

$$JS\left(f(x)\|g(x)\right) = \frac{1}{2}[KL\left(f(x)\|g(x)\right) + KL\left(g(x)\|f(x)\right)]. \tag{6.10}$$

Since relative entropy allows us to compare two functions, it was first applied to signal processing. The greater the relative entropy, the more different the signals, and vice versa. Later, information processing researchers adopted relative entropy to measure the distance between pieces of information. For example, a plagiarized article will share many of the same words with the original, so the two will exhibit a low relative entropy. In the same spirit, Google's automatic question-answering system uses the Jensen-Shannon divergence to measure the similarity between two answers.

Relative entropy can also determine whether the probability distribution of commonly used words is the same between different texts. This result leads to one of the most important concepts in information retrieval, text frequency-inverse document frequency (TF-IDF). TF-IDF can evaluate the relevance of search results, as well as categorize news articles. We will cover the specifics of TF-IDF in later chapters.

6.5 Summary

Entropy, conditional entropy, and relative entropy, are closely related to the statistical language model. When we discussed the statistical language model in Chapter 2, we did not disclose how to assess the quality of a given model. At that time, we did not have these three concepts at hand.

Naturally, any good model will produce low error rates, whether it be applied to machine translation, speech recognition, or other fields. Today's natural language processing systems *do* have low error rates, but error rate is neither direct nor convenient for researchers to evaluate. When Jelinek and peers studied statistical language models, there was no gold standard against which to compare their work. Simply put, there did not exist any decent machine translation or speech recognition system to provide a baseline error rate. Rather, the point of a language model is to predict the meaning of text in context, so the better the model, the lower its uncertainty.

Information entropy is exactly this measurement of uncertainty. Since language is contextual, we use conditional entropy for high-level language models. If we take into account the deviations between training corpus and real-world text, we add in relative entropy. Based on condition entropy and relative entropy, Jelinek derived a concept called perplexity, quantifying the quality of a language model. A language model's perplexity has a definite physical meaning: given some existing context (e.g., parts of a sentence), a model with higher perplexity will allow more words to fill in the gaps, and vice versa. Thus, the lower a model's perplexity, the more certain we are of each word, and the better the model.

In his 1989 paper about the Sphinx speech recognition system, Dr. Kai-Fu Lee used perplexity to demonstrate the improvements made by his language models. Without any model, a sample sentence could be filled by 997 possible words, giving a perplexity of 997. After a bigram language model was introduced, the perplexity immediately dropped to 60. After switching to probabilistic language model instead of counting pairs of words, Lee further reduced the perplexity to a mere twenty.

Readers with interest in information theory and ample mathematical background may reference the "Elements of Information Theory," written by Professor Thomas Cover from Stanford University. Professor Cover was one of the most authoritative experts on information theory.

Bibliography

1. Cover, Thomas M. and Thomas, Joy A. Elements of Information Theory, New York: Wiley, 1991. ISBN 0–471-06259-6.

2. Lee, Kai-Fu, Automatic Speech Recognition: The Development of the SPHINX System, Springer, 1989.

3. Gale, W., K. Church, and D. Yarowsky. "A Method for Disambiguating Word Senses in a Large Corpus." Computers and the Humanities. 26, pp. 415–439, 1992.

Chapter 7

Jelinek and modern language processing

I dedicate this chapter to Dr. Frederick Jelinek.
18 November 1932 — 14 September 2010

When I first published the "Beauty of Mathematics" articles on Google China's blog, I wrote about some of the most successful scholars who applied mathematics to natural language processing. Admittedly, those stories served to pique the readers' interests, but their fundamental purpose was not to tell entertaining tales or gossip. Rather, I wanted to introduce a group of masters to young people who are interested in similar work, so that they can learn from the masters' ways of thinking.

In our materialistic societies, the youth are under a lot of stress to succeed, and the successful are lonely in their eminence. This was very much the situation in Romain Rolland's post-war France. Rolland wrote *Three Biographies* (Beethoven, Michelangelo, and Tolstoy) for young men who wished to pursue

FIGURE 7.1: Frederick Jelinek.

noble causes rather than material wealth,* to breathe from the air of giants. Today, I hope to introduce a similar group of masters to aspiring students. We start with Frederick Jelinek.

Readers who have sequentially read up to this chapter may recall the name Jelinek from previous sections. Indeed, Jelinek is indispensable from modern natural language processing. However, here I focus not on his contributions towards science, but on his life as an ordinary person. These stories I tell come from either my own experience, or from his personal recounting to me.

7.1   Early life

Frederick Jelinek (hereafter referred to as "Fred"; see Figure 7.1) was born to a wealthy Jewish family in Kladno, Czech Republic.† His father was a dentist, and in Jewish tradition, Fred's parents highly valued his education. They planned to send him to a private school in Britain, and even hired a German governess to perfect his German.

Unfortunately, the Second World War shattered their dreams. The family was driven out of their homes and they moved to Prague. Fred's father died in a concentration camp, so Fred played on the streets all the long, leaving school far behind. After the Second World War had ended, Fred returned to elementary school, but his grades were a mess. He started with Ds and quickly caught up to his classmates, but not once did he receive an A in elementary school.

In 1946, the Communist Party seized control of Czechoslovakia, and Fred's mother knew better than to stay, from her late husband's unfortunate fate. She decided to immigrate to the United States. There, the family was destitute; their only income was from the pastries that their mother sold. As a teenager, Fred worked in the factories to earn some extra money for the family. Consequently, Fred had little time to spend at school or at home. Before college, he could only study for a fraction of the time good students do nowadays.

Personally, I attended primary school during the Cultural Revolution and college during the 1980s reform era, so I also spent less than half of my time studying. Therefore, neither of us supports the status quo that students attend long days of school to score well on exams. Whenever Fred and I discussed our own education, we agreed on these points.

1. Primary and secondary school students should not spend all their time in class. The social interactions, life skills, and aspirations they acquire will define them for a lifetime.

Life of Beethoven, Life of Michelangelo, and *Life of Tolstoy.* Rolland was awarded the 1915 Nobel Prize in Literature for his idealism and sympathy towards humankind.

†A city some 25 kilometers from Prague, the Czech capital.

2. Gaining an advantage in primary or secondary school requires a lot more time than in college, because in college, the mind is more mature and more capable of understanding knowledge. For example, a high schooler may spend 500 hours learning material that a college student can learn in only 100 hours. Thus, early advantages are less important at the college level.

3. Learning is a lifelong process. Some Asian students excel in secondary school and gain admittance into elite colleges, where they perform significantly worse than their American peers. The former were not driven enough to continue their studies, while the latter studied for personal interest and continue to study out of innate motivation.

4. We can learn textbook material before our peers do, or we can learn it after them, but we cannot make up the critical stages of growth from child to adult. Today, in China's top secondary schools, students probably spend 99% more time studying than I did, and even more than Jelinek. Regrettably, they probably have fewer academic accomplishments than I, not to mention Jelinek. This is quite a misstep in education.

At ten years old, Jelinek's first dream was to become a lawyer so that he could prevent miscarriage of justice, as had befallen his father. Upon arriving in the United States, however, he realized that his thick foreign accent would make court defense nigh impossible. His second goal was to become a doctor, like father like son. He hoped to attend Harvard Medical School, but he could not afford eight years of tuition (four years of undergraduate education, and four years of medical school). Fortunately, at this time the Massachusetts Institute of Technology (MIT) granted him a full scholarship, intended for Eastern European immigrants, so Jelinek set off to study electrical engineering at MIT. Throughout this journey, it seems that Jelinek's vision of his future was constantly in flux. However, he had an unwavering determination to succeed through hard work.

At MIT, Jelinek met world-class researchers, including Dr. Shannon, the founder of information theory, and renowned linguist, Roman Jakobson, who developed the six functions of communication.* Later, his wife, Milena Jelinek, immigrated to the United States to study at Harvard University, so Fred often listened in on lectures with his wife. There, he frequently attended the class of Noam Chomsky, the father of modern linguistics.

These three masters heavily influenced the direction of Jelinek's later research, to solve language problems with information theory. I have always believed that if anyone wants to become a world-class researcher in his own field, then he must be surrounded by a cohort of distinguished mentors. Jelinek was

---

*The Jakobson model is presented in Chapter 3.

fortunate to receive these masters' advice as a young student, so his work far surpassed that of his peers later in life.

After receiving his Ph.D. from MIT, Fred taught for a year at Harvard, then became a professor at Cornell University. During his job search, he had exchanged ideas with linguist Charles Hackott, who promised him that they would collaborate on solving linguistics problems through information theory. Once Jelinek arrived, however, Hackott told Jelinek that he was no longer interested in solving language with information theory; he had turned to writing operas. Thus sparked Jelinek's distaste for linguists.

Later at IBM, Jelinek developed the opinion that linguists spoke in fancy rhetoric without making progress on the problems at hand, so he famously quipped, "Every time I fire a linguist, the performance of the speech recognizer goes up." This phrase was widely circulated in the natural language processing industry and lives to this day.

7.2   From Watergate to Monica Lewinsky

"From Watergate to Monica Lewinsky" was not a title chosen for comic effect alone; it was actually one of Jelinek's sections in his 1999 ICASSP* report to the general assembly.[†] The Watergate incident happened to fall just as statistical speech recognition took off, and President Clinton had recently been impeached for the Lewinsky affair, one year before the conference.

After ten years of rigorous research at Cornell, Jelinek discovered the key to natural language processing. In 1972, he left on sabbatical for IBM Watson Laboratory, where he unintentionally became head of the Speech Recognition group. Two years later, he chose to permanently leave Cornell for IBM. There, Jelinek formed an impressive team of researchers: his partner L. Bahl; James and Janet Baker, founders of Dragon speech recognition company; twin brothers S. Della Pietra and V. Della Pietra, who developed the maximum entropy iteration algorithm; J. Cocke and J. Raviv, co-authors of the BCJR algorithm; and Peter Brown, the first to propose a statistical model for machine translation. Even the most junior of the members, John Lafferty, became an accomplished scholar in his later years.

The 1970s IBM was much like the 1990s Microsoft or present day Google, where outstanding scientists are given the freedom and funding to pursue their personal research interests. In that environment, Jelinek and others proposed the framework for statistical speech recognition. Before Jelinek, speech recognition was treated as a problem of classical artificial intelligence problems or pattern matching. Jelinek considered speech recognition a communication problem, which could be solved by two hidden Markov models (acoustic and

---

*International Conference on Acoustic, Speech and Signal Processing.
[†]http://icassp99.asu.edu/technical/plenary/jelinek.html.

linguistic). His framework had far-reaching implications: it not only made speech recognition computationally feasible, but also set the foundation for all natural language processing today. For this innovation, Jelinek was elected to the National Academy of Engineering and named one of *Technology* magazine's top 100 inventors of the 20th century.

While Jelinek was intellectually primed to develop his statistical framework, he was also born into the right era for such a discovery. Jelinek's predecessors, Shannon et al., faced two insurmountable obstacles in applying statistical methods to natural language. First, they lacked the computational resources required to tackle big data, and second, they lacked a large, machine-readable corpus for natural language research. In the end, Shannon and his peers were forced to give up.

When Jelinek joined IBM in the 1970s, the computing power at his disposal was still incomparable to what is widely available today, but it was sufficient for many of his experiments. Thus, his remaining problem was a machine-readable corpus. Today, any decent web crawler can unearth copious amounts of data, but back then, there were no web pages. In fact, most publications did not have electronic records, and if they did, they were held by different publishers. Fortunately, there existed a global telecommunications network, known as telex. As a result, IBM researchers embarked on the natural language problem using telex records.

In retrospect, it was inevitable that IBM would solve the natural language problem in the 1970s. First, only IBM had sufficient computing power and digital data in that era. Furthermore, Jelinek and peers had spent a decade working on the theoretical side of this field. When he took sabbatical at IBM, the opportunities for collaboration were self-evident. Finally, Watson Jr. led the IBM business to its peak in the 1970s, so IBM had the resources to invest heavily in basic research. If young people at that time could connect the dots and had sufficient mathematical background (criteria for hiring), they would have seen that IBM led to a promising future.

Another major contribution of Jelinek, Bahl, Cooke, and Raviv was the BCJR algorithm, one of the two most widely used algorithms in telecommunications today (the other is the Viterbi algorithm). Today, IBM lists this algorithm as one of its greatest contributions to humanity on the walls of California's Almaden Research Lab. Interestingly enough, this algorithm was only widely adopted twenty years after its invention. Unfortunately, all four of BCJR's inventors had left IBM by then. When researchers began to implement this algorithm, IBM's communication department had to hire an expert from Stanford University to explain their own algorithm. Upon arrival, that expert was astonished by the volume of achievements IBM had accomplished.

At the 1999 ICASSP conference held in Phoenix, Jelinek gave a report entitled, "From Watergate to Monica Lewinsky," encapsulating 30 years of progress in speech recognition. He gave a summary of the work completed at IBM and later, Johns Hopkins, the latter of which included my own research.

Many years later, I had a conversation with Alfred Spector* about why IBM, of all companies, was able to solve the speech recognition problem. Unlike AT&T, Bell Labs, or Carnegie Mellon University, IBM had no previous experience dealing with speech recognition. Spector believed that with no previous research, IBM started from a clean slate, uninfluenced by the predominant ways of thinking at the time. However, this is only one factor to IBM's success, and I emphasize that most historical developments do not arise by chance alone. IBM seemingly stumbled upon statistical language processing, but at that time, only IBM was ready for such a discovery. It was the only place with enough computing power and resources to assemble the world's brightest minds.

7.3   An old man's miracle

Readers of *On Top of Tides* may recall that IBM experienced some of its hardest times during the late 1980s and early 1990s, when former CEO Gerstner heavily cut the research budget. As fate would have it, speech recognition and natural language processing were also on the list of budget cuts, so Jelinek and some of IBM's most distinguished scientists left IBM in the early 1990s. Many headed for Wall Street, where they became billionaires, but Jelinek would take a different path. At that time, Jelinek had reached the conventional age to retire, and he had accumulated enough wealth to live out old age in comfort. Though most people in his position may have chosen retirement, Jelinek had been a busy man his whole life and his scholarly spirit would not rest. So in 1994, he founded the Center for Language and Speech Processing (CLSP) at Johns Hopkins University.

Prior to Jelinek's arrival, Johns Hopkins University was still well regarded for its medical school, but it had fallen far behind in the engineering disciplines. After World War II, Johns Hopkins had little hope of catching up to the Massachusetts Institute of Technology or the California Institute of Technology. At Johns Hopkins, no one even studied natural language processing or speech recognition, both relatively new fields at the time.

From this disadvantaged situation, Jelinek worked his magic, building the CLSP from the ground up. By its second or third year, the CLSP had become a world-class research center. To this end, Jelinek carried out two major tasks and two minor tasks. Most significantly, Jelinek acquired extensive funding from the US government, and he used this money to invite 20-30 of the world's top scientists to collaborate at the CLSP each summer. With these experts in attendance, the CLSP became one of the world's major centers for natural language processing.

In addition to his resources, Jelinek recruited a group of promising young scholars, including Yarowsky (later famous for his natural language processing

---

*He has served as the VP of research at Google and IBM.

work) and Blair (eBay's vice president of research). Furthermore, Jelinek used his influence to send his students on internships at the best companies around the world. Through these interns, the CLSP developed a reputation for nurturing outstanding talent. More than ten years later, the United States government decided to establish a Center for Excellence to address national security needs. Johns Hopkins outbid its longstanding rivals, MIT and Carnegie Mellon University, to secure its position as one of the world's top research universities.

As an advisor, Jelinek was very rigorous and demanding towards his students. He had a notoriously high rejection rate, and those who remained were kept for a long time before they were allowed to graduate. In exchange, Jelinek leveraged all of his influence to facilitate his students' learning and careers. From a student's first day at Johns Hopkins to his last day before graduation, tuition and living expenses were fully paid for. To improve the English proficiency of foreign students, Jelinek used his own funding to hire private English tutors. Jelinek also ensured that each doctoral student experienced at least one internship at a major company before graduating. Those who earned their Ph.D. from Jelinek found jobs at prestigious laboratories, such as IBM, Microsoft, AT&T, and Google.

Jelinek created a generation of talented scientists, including his own students, his past subordinates, and younger students who followed in his footsteps. Of those include Peter Norvig, head of Google research, and Fernando Pereira. These people form a school of thought across major universities and companies around the world, and Jelinek stood at the head of this school.

Academically, Professor Jelinek's most valuable gift to me was to augment my academic horizons with his own research experience. Most often, he told me which methods I *shouldn't* try. Jelinek's advice was similar to that of Warren Buffet's to his lunch guests.* Buffet tells his guests that they are smart; they can figure out what will work. Buffet tells them what he has seen to fail. In the same vein, Jelinek diverted me from unfruitful paths that he had already explored with his colleagues at IBM. He believed that I was more capable than he is at making new discoveries. Under his guidance, I was saved from wasted effort and I could focus on making progress. This way of thinking has benefitted me for a lifetime.

Jelinek lived a frugal life. For twenty years he drove an old Toyota, more dingy than any of his students' cars. Each year, he invited his students and professors from home and abroad to his house. Graduates often came back for this circumstance. There, we stopped talking about academic issues and opted for recent movies (his wife was a professor of film at Columbia University) or the latest scientific gossip. We learned why Professor Thomas Cover of Stanford University was unwelcome at all of Las Vegas's gambling houses and more.

---

*Each year, Buffet eats lunch with one investor, selected by auction. He donates the highest bid to charity.

At these gatherings, the only deficiency was the food. Jelinek would always serve carrots and celery sticks. On one occasion, he paid Professor Miller (in the same department) to host the party at his house instead. Miller hired a professional chef who cooked a delectable meal, complete with wine. From thenceforth, the annual party was transferred to the Miller family.

Besides these parties, let me share a few other fond memories of Jelinek. Jelinek's wife, Milena, studied film, so Jelinek started watching Chinese films very early on. China's early international films almost always featured the Singaporean-Chinese actress, Gong Li, so Jelinek always wondered why such a large country only had a single actress. Jelinek's first understanding of China consisted of Tsinghua University (from which I graduated) and Tsingtao beer. He often confused the two names—once, to his chagrin, in front of Professor Pascale Fung from the Hong Kong University of Science and Technology.

Jelinek was straightforward and to the point when he spoke. If we discussed academics with him, we took extra care to be rigorous, otherwise he would catch us in our mistakes. In addition to his aforementioned remark about linguists, Jelinek was known for his biting words towards many world-class researchers. When someone did succeed, however, Jelinek was equally lavish with his praise. In 1999, I won the Best Paper Award at Eurospeech, the European natural language processing conference. Upon my return to the lab, Jelinek greeted me with, "we are so proud of you." Throughout his 40-year academic career, Jelinek truly offended few. I attribute his rapport with the scientific community not only to his accomplishments, but also to his virtuous character.

As I have mentioned, Jelinek was a man who would not settle down. I often saw him working overtime in the lab on weekends. He was still sharp-minded into his seventies and came to work on time, every morning. On September 14, 2010, Jelinek arrived at his office as usual, but unfortunately, he died of a heart attack at his desk. I was sad and shocked when I heard the news because he was fine when I visited him at Johns Hopkins, just months earlier. At an age when most retire, Jelinek founded the world's largest natural language processing center of its era. Many years ago, Jelinek and I discussed that learning is the journey of a lifetime. For Jelinek, it truly was, until the last day of his life.

Many of Jelinek's students and friends later came to work at Google. Together, they donated a large sum of money to Johns Hopkins and dedicated the Jelinek scholarship, to which interested students may apply.*

---

*http://eng.jhu.edu/wse/Fred_Jelinek/memory/jelinek-fellowship.

Chapter 8

Boolean algebra and search engines

In the next few chapters, we tackle web search and related technologies. When I first posted this series on Google China blog, some readers were disappointed that I only covered the main ideas behind search without delving into details. This time, I am afraid I will have to disappoint some readers again, because our discussion remains on the surface level for several reasons.

First and foremost, this book's intended audience is the general public, not just employees of search companies. To the layperson, understanding the role of mathematics in engineering is more important than explicating every single algorithm hidden behind a product. Furthermore, the bases of technology are twofold: theory and implementation. This book aims to ground the reader in the theories behind innovations, rather than provide their code documentation. A lot of search technology will be out of date in a few years, replaced by newer, better versions, but the theory behind them does not change. Those who continually pursue the latest implementations lead hard lives, because they must renew their knowledge every few years. Only those who understand the nature and essence of search can remain relevant throughout their careers.

Finally, many people who wish to know the underlying implementation of Google's search engine seek a shortcut to success. There is no such shortcut to a job well done; we cannot bypass 10,000 hours of professional training and hard work. To construct a quality search engine, the most basic requirement is to analyze 10-20 bad search results every day. Over time, you will develop a sense for areas of improvement. When I worked on Google's search team, I looked at far more results per day, and Amit Singhal, Google's chief technology officer for search quality, still surveys poor search results as part of his job. However, many engineers in the United States and China cannot commit to this dirty work. They still hope for an algorithm that will solve all search problems, and such an algorithm is unrealistic.

Now, we return to the technical aspect of search engines. The basic recipe is simple: automatically download as many web pages as possible; generate an efficient and useful method of indexing; and sort the web pages in order of relevancy, by some fair and accurate metric.

When I left Google to become a VP at Tencent, I organized all of Soso's* functionalities into three modules—download, index, and ranking. These three

---

*Soso was formerly Tencent's flagship search engine.

formed the backbone of search and corresponded to the theory behind search. All search applications were built atop these modules, and those applications constituted the specific implementations. To improve the quality of Tencent's myriad search applications, we first unified them through these fundamentals, which we subsequently ensured were of the utmost quality. Only then could we trust that the company's search was sound. If we had instead worked from tweaking each search application separately, we would have been redesigning skyscrapers atop shaky sand castles. Likewise, this book introduces search from the ground up—we focus less on the specific applications and more on the foundations.

8.1   Boolean algebra

There is no simpler number system in the world than binary. Binary has two numbers, 0 and 1. From a purely mathematical standpoint, it makes more sense to count in binary than in decimal. Alas, humans have ten fingers, so evolution and history dictate that we count in base ten, not binary or octal.

Though decimal dominates most transactions between people, binary was also invented in the dawn of human civilization. It can be said that the Chinese belief in yin and yang was the earliest manifestation of binary thinking. As a counting system, binary was first used by Indian scholars between the second and fifth centuries BC, although they did not have the concepts of 0 and 1. By the 17th century, however, humankind invented binary in its modern form. German mathematician Gottfried Leibniz refined the theory of binary and introduced 0 and 1, arriving at the system that today's computer scientists know and love.

In addition to representing numbers, binary can also represent a logical "yes" or "no." This second feature provides great value to search engine indexing. Just as addition and subtraction can act on decimal numbers, so too boolean operations act on logical values. Given the simplicity of binary numbers, we might guess that boolean operations are equally easy to understand. Every modern search engine claims to be intelligent and optimized (really feats in advertising), but each relies on the simple idea of boolean algebra.

Before we introduce the mathematics of boolean algebra, let us dive into the ironic history behind its inventor. During the 19th century, George Boole was a high school math teacher in the United Kingdom. He founded a secondary school himself and later became a professor in Cork, Ireland. No one ever considered him a mathematician throughout his lifetime, although he did publish several papers in the *Cambridge Mathematics Journal*.* In his spare time, Boole enjoyed reading mathematics papers and contemplating open problems.

---

*These misconceptions are far from uncommon. Another famous British physicist, James Joule, was never considered a scientist his whole life. Although he was a member of the Royal Academy of Sciences, his official job was a beer brewer.

In 1854, he published *An Investigation of the Laws of Thought, on which are founded the Mathematical Theories of Logic and Probabilities*, which united mathematics and logic for the first time. Before then, mathematics and logic were considered two distinct disciplines, and to this day, UNESCO strictly lists them as separate.

The actual mathematics behind boolean algebra should be familiar to the reader. There are two basic units: 1 represents TRUE and 0 represents FALSE. These units can be manipulated by three operations: AND, OR, and NOT, which mean the same as in English. It was later discovered that all operations can be reduced to the AND-NOT operator. Tables 8.1, 8.2 and 8.3 illustrate these boolean operators.

TABLE 8.1: If one operand is a 0, then the AND operator must return a 0. If both operands are 1, then the AND computes a 1. For example, the statement "the sun rises from the West" is false, and "water can flow" is true. Therefore, "the sun rises from the West AND water can flow" is false.

AND	1	0
1	1	0
0	0	0

TABLE 8.2: If any operand is a 1, then the OR operator always returns a 1. If both operands are 0, then the result is 0. For example, "Trump won the 2016 US presidential election" is true, while "Clinton won the 2016 US presidential election" is false. So "Trump OR Clinton won the 2016 election" is true.

OR	1	0
1	1	1
0	1	0

TABLE 8.3: The NOT operator flips 0 to 1 and 1 to 0. For example, suppose that "flowers do smell good" is true. Then the statement "flowers do NOT smell good" must be false.

NOT	
1	0
0	1

Readers may wonder how this elementary number system relates to modern technology at all. Boole's contemporaries certainly asked the same question. In fact, no one had created any nontrivial system using boolean algebra for 80 years after its inception. It was only until 1938, when Shannon's master thesis applied boolean algebra to digital circuits, that the field really took off. All mathematical operations—addition, multiplication, exponentiation, etc.—can be reduced to a series of boolean operations. This property of boolean algebra allowed humans to implement computers from many little switches. As we discussed in the first chapter, mathematics is a process of abstraction that appears to travel farther and farther from everyday life, but eventually it finds a place, just as boolean algebra found a home in our digital revolution.

Now let us examine the relationship between boolean algebra and search. When a user enters in a keyword, the search engine examines whether each document contains that keyword. Documents that contain the keyword are assigned 1, or TRUE, while documents that do not are assigned 0, or FALSE.

For example, to search for literature about atomic energy, but not about atomic bombs, you can write a query string,

atomic AND energy AND (NOT bomb).

This query string retains all documents related to atomic energy and discards those related to atomic bombs. We can evaluate this query on each document, according to the truth tables provided for each boolean operation. Thus, we have combined logic and mathematics to determine which documents are relevant.

The significance of boolean algebra to mathematics is akin to that of quantum mechanics to physics. Modern physics show that our world is made up of finite elementary particles[*], much fewer than a googol (10^{100})[†] of them in fact. Similarly, the world of boolean algebra quantizes all values either as true or false. Quantum mechanics are often described as counterintuitive; boolean operations, including AND and NOT, are quite different from the traditional algebra we learn in secondary school. Our seemingly continuous world is actually very discrete, both in physics and in computer science.

8.2   Indexing

Consider the numbers for a second. A search engine can sift through hundreds of millions of search results in tenths of a second, to give you the most

---

[*]According to http://www.universetoday.com/36302/atoms-in-the-universe/, the number of particles in the universe is between 10^{78} and 10^{82}, counting from the smallest basic particles (quarks, electrons, photons). When taking into consideration dark matter and dark energy, this number should not exceed 10^{86}.

[†]Google's name is derived from this quantity, which represents an enormous amount of indexing.

quality results on the first page—isn't that impressive? Obviously the computer cannot *read* all the articles on the spot; even with Google's computing resources, that would be too slow. The trick is to build an index, the same sort that appears at the end of textbooks or in libraries.

One of Google's interview questions for the product manager position asks: "how do you explain a search engine to your grandmother?" Most candidates come from search companies and immediately dive into the technical details. Those solutions mostly fail. A good answer might include an analogy to a library catalog's index cards. Each website is a book in the library. If a customer requests the best books about Bengal cats, we cannot scour the bookshelves, read them all, and deliver the relevant ones. Instead, we search the index cards and directly retrieve the desired books from the shelves.

Of course, we cannot apply boolean algebra to index cards, but as information retrieval entered the digital age, physical cards became database entries. Database query languages (e.g., SQL) support a variety of complex logical combinations, but the underlying rationale is based on boolean operations. Early search engines required that you specify the exact boolean formula (with ANDs and ORs), but modern search engines are smarter. They automatically convert user queries to such formulas.

Boolean algebra would not be complete for search engines if we did not include a smart representation for keywords and documents. The easiest approach is to use a long binary number, indicating whether a given keyword appears in each document. There are as many digits as documents. A 1 at location i indicates that document i has the specified keyword, while a 0 at that position means the corresponding document does not have that keyword. For example, in the case of "atomic," the corresponding number might be 0100100011000001..., so documents 2,5,9,10,16, etc. are relevant to atomic-something. Similarly, suppose the binary number 0010100110000001... corresponds to "energy." If we AND those two numbers, the resultant 0000100000000001... indicates that documents 5 and 16 are related to atomic energy, which is our desired topic.

Computers are very, *very* fast at boolean operations—they are what computers were designed to do. Even the cheapest microcomputer can perform 32-bit boolean operations in a single clock cycle, with the processor spinning billions of cycles per second. Furthermore, since most entries in this binary number are 0, we only need to record the indices at which they are 1. As a result, these entries stack up to a large table: each row corresponds to a single keyword, followed by a set of numbers that represent relevant documents containing the keyword.

To a search engine, every website is a document. As one might expect, there is an enormous number of web sites on the Internet, and within them, the number of keywords is on the order of trillions. Historical search engines such as AltaVista were limited by the computer speeds and capacities of their times, so they could only index the most important keywords. To date, many academic journals require that authors provide 3-5 keywords, so their articles are more searchable. However, representing an entire article by a few words is bound to overlook some aspects, so the largest search engines actually index all the words in a document. Unfortunately, it is a nontrivial engineering feat to read every single document on the Internet.

To illustrate, we consider this (conservative) hypothetical case. Suppose there are 10 billion meaningful pages on the Internet* and our dictionary is 300,000 words large. Then the index will contain at least 10 billion × 300,000 = 3,000 trillion bits. These data are naturally sparse, so we can estimate a compression rate of at least 100:1, yielding 30 trillion bits (over 3,000 gigabytes). The sparse representations require slightly more bits per location, and to facilitate ranking the search results, we should include auxiliary data, such as word frequencies or locations. Just the memory required to store this index exceeds that of any single server.

Therefore, the index must be stored in a distributed fashion, across many servers. A common practice is to split the index into different shards, depending on the page number. Each time a query arrives, it is dispatched to multiple servers that execute in parallel. When those servers are finished, they send their results back to the master server, which consolidates the information and returns the result to the user.

With the increase in Internet access, especially in the Web 2.0 era, users are generating more and more data every day. Even companies like Google, with seemingly unlimited computing power, are feeling the pressure of more data. As a result, indices are split into many tiers, depending on their importance, quality, and frequency of visits. Commonly used indices require fast access, high redundancy, and frequent updates. On the other hand, uncommon indices have lower performance requirements. In the grand scheme, however, no matter the technical challenges, the theory behind search engines is still simple— boolean algebra.

8.3   Summary

Boolean algebra is simple, but it holds great importance in modern mathematics and computer science. Not only did it merge logic and mathematics, but it also provided a new perspective on our world and opened up the digital age. To end this chapter, we borrow some of Newton's words:

Truth is ever to be found in simplicity, and not in the multiplicity and confusion of things.

---

*The actual number is much greater.

Chapter 9

Graph theory and web crawlers

Discrete mathematics is an important branch of contemporary mathematics and the foundation of theoretical computer science. Under its umbrella are formal logic, set theory, graph theory, and abstract algebra. Logic is strongly tied to boolean algebra, as previously discussed. If you search "discrete mathematics" on Google Trends, you will find that the states most interested in this field are Maryland, Indiana, New York, Virginia, California, Massachusetts and New Jersey—all locations with well-established universities.

In the previous chapter, we described the main ideas behind a search index, which knows the websites related to each potential keyword—but how do we obtain all the websites in the first place? Here, we will introduce graph theory and its relationship to the Internet's automatic download tools, web crawlers. Graph traversal algorithms help us walk the entire Internet.

9.1 Graph theory

The origins of graph theory can be traced to Leonhard Euler. In 1736, Euler took a trip to Königsberg, Prussia*, where the locals had a peculiar pastime. Königsberg had seven bridges, as depicted in Figure 9.1. The goal was to walk through each bridge exactly once and return to the original point of departure, but no one had ever succeeded. Euler proved that succeeding was actually *impossible* and published a paper to prove this case. People consider this paper the inception of graph theory, and we will describe Euler's proof in the extended reading.

Euler's paper described a structure composed of vertices and edges that connect those vertices. For example, if we consider American cities as vertices and roads as edges, then the Interstate Highway System is such a graph. There are many algorithms regarding graphs, but some of the most important ones solve the graph traversal problem: given a graph, how can we visit all of its vertices by traveling on its edges?

Consider the Interstate Highway System, as depicted in Figure 9.2. There are cities directly reachable from New York City, including Boston and Baltimore (of course, there are roads and cities not drawn). After we visit

---

*Hometown of the philosopher Kant and now known as Kaliningrad, Russia.

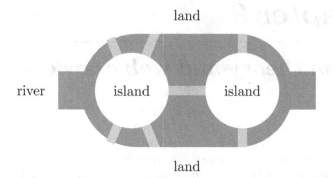

FIGURE 9.1: Seven bridges of Königsberg.

Boston or Baltimore, we are able to travel to more places, including Miami, Indianapolis, and Cleveland. Then we can reach Atlanta, Nashville, Chicago, etc., until we see all the cities in the United States. This method of graph traversal is known as breadth first search (BFS), since it explores all connections as broadly as possible, instead of selecting a single path to fully explore.

   Another strategy is to start at New York City and keep traveling along the road until there are no more new cities, at which point we backtrack and take all the forks in the road. For example, we could travel from Boston to Baltimore, then south to Miami, Atlanta, Nashville, and Indianapolis. At Indianapolis, we notice that there are no unseen cities, so we return to Nashville and head north towards Chicago. This strategy is known as depth first search (DFS) since we pick a path and thoroughly explore it to its end.

FIGURE 9.2: American highway system as a graph.

In both cases, we require some bookkeeping to make sure that we do not visit the same city twice (wasted effort) and that we do not skip any cities. Both methods ultimately guarantee that we will visit all the cities on the map.

9.2 Web crawlers

Now that we understand the basics of graph theory, we can uncover the relationship between graph traversal algorithms and web crawlers. Although the Internet is an enormous, complicated place, it too can be represented as a graph. Websites are vertices, and hyperlinks between them are edges. Most Internet users probably know that behind every clickable URL is a hyperlink target, which will take you to another page on the web. Those links are specified within a website's source code, so if we download a web page, we can use graph search to explore all the outgoing links from that page. Programs that automatically explore the Internet are known as web crawlers, or "bots." The first web crawler was invented by an MIT student, Matthew Gray, in 1993. He named his program the "WWW Wanderer." Modern web crawlers are much more complicated than Gray's contraption, but the underlying ideas have not changed.

Recall that a web crawler's purpose is to download the entire Internet. That's a big task! But we can start small. We first download a single web page and retrieve all outgoing hyperlinks that are directly available. If we start at Google.com, then we discover links to Gmail and Google Images within our first pass. Next, we download all of the pages we just found and retrieve their outgoing hyperlinks. If a computer keeps working at it, it will eventually download the entire reachable Internet. In addition to downloading as much as it can, the web crawler needs to keep track of where it has been, to avoid duplicates. People often use a hash table for this purpose.

In our theoretical discussion of web crawlers, we have been avoiding one key technical challenge: the Internet's size. For reference, Google had indexed one trillion websites by 2013, of which 10 billion entries were updated frequently. Even if we could download one web page every second, it would require 317 years to download 10 billion of them and 32,000 years to download all trillion. Human records have only been around for one sixth of that time! So a commercial web crawler requires tens of thousands of servers, all connected to a high-speed network. This array of servers and their coordination belong to the art of network architecture.

9.3 Extended reading: Two topics in graph theory

9.3.1 Euler's proof of the Königsberg bridges

Recall the Königsberg bridges in Figure 9.1. If we take each landmass as a vertex and the bridges as edges, we construct the graph in Figure 9.3.

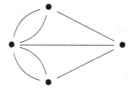

FIGURE 9.3: Abstraction of the Königsberg bridges.

For each vertex in the graph, we define the number of outgoing edges as its degree.

Theorem 1. *Given a graph, if there exists a path that traverses all the nodes and edges without repetition and returns to the starting node, then the degree of each vertex must be even.*

Proof. Suppose there exists such a path. Then for each vertex, every edge leading in has a counterpart leading out. The number of times we enter and exit each vertex is the same, so edges come in pairs. Therefore, the degree of each vertex is even. ■

In Figure 9.3, we see that some vertices have odd degrees, so some vertices cannot be visited without repeating an edge.

9.3.2   The engineering of a web crawler

One of my favorite Google interview questions is, "how do you construct a web crawler?" I asked this question so often that some interviewees anticipated it, but this question still selected the strong candidates from the weak. It assesses all aspects of a candidate's computer science theory background, algorithmic knowledge, and practical engineering experience. Its versatility lies in the lack of a correct answer; there are only good answers and bad answers, with boundless depths to explore. A strong candidate can answer this question without experience building web crawlers, but a third-rate engineer working on web crawlers will not be able to provide all the high-level details.

There are many engineering considerations behind a web crawler, the first of which is the primary algorithm—BFS or DFS?

In theory, if we do not consider time constraints, both algorithms can "crawl" the entire Internet in roughly the same time.* However, real world constraints are unavoidable, so a web crawler's more realistic goal is to download the most important pages within given time limits. By common sense, each website's most important page is its homepage. In extreme cases, if the web crawler is

---

*If there are V vertices and E edges, then both algorithms take $O(V + E)$ to traverse the whole graph.

small, then it should only download the homepage. Larger web crawlers can then focus on pages directly linked from the homepage, as a web designer deemed those pages important. Under these premises, it would appear that BFS is more suitable than DFS. Although practical implications prevent a direct BFS implementation, the underlying heuristics of which pages to download first essentially follow BFS.

Though BFS inspires the big ideas, DFS also plays a part in web crawling. Before a web crawler can download a web page, its server must undergo a "handshake" process to establish communications. Performing the handshake requires nontrivial overhead time, and too many handshakes can actually reduce download efficiency. Fortunately, since the web crawler is composed of many distributed servers, each website can be allocated to one (or several) servers. These servers will download an entire website's data in one go, rather than downloading 5% at a time and performing a new handshake each time. By following a DFS approach for individual websites, we can avoid excess handshakes.

On a whole, web traversal cannot be solved by purely BFS or DFS. Practical implementations generally involve a scheduler, which determines which web pages to download next. Pages that have been discovered, but not downloaded, are stored in a priority queue by URL. In the big picture, however, traversing the Internet is still more similar to BFS, so BFS more heavily determines the priority of websites.

Now we have selected an algorithm, our next order of business is to actually extract all outgoing links from a web page. In the early days of the Internet, this task was easy, since the Internet was written in pure HTML. All URLs would be delimited by specific tags, indicating a hyperlink, so those could be directly parsed. Nowadays, URL extraction is a bit more complicated. Many websites are dynamically generated using scripting languages like JavaScript. Often, the source code does not directly offer up hyperlink addresses; rather, the URLs are inserted after running the contents of the page.

As web crawlers become more complex, they often simulate browsers to run web pages and generate the resulting data. Unfortunately, some websites are poorly written and their scripts are difficult to simulate. Those websites are still accessible through regular browsers, so they exist to the public, but they pose challenges to the web crawler. Given the close relationship between browsers and the Internet, browser engineers are often called upon to help design better web crawlers. Sadly, there are not enough quality browser engineers to be found. Therefore, if you find a website that obviously exists without showing up in the search engine, then the search engine may have failed to parse the non-standard scripts on that page.

The last challenge I often discuss is the matter of recording seen URLs. On the Internet, there may be many ways to reach the same website. In graph theory terms, the website vertex has many incoming edges. So during graph search, we may potentially visit that website many times. To prevent duplicated effort, we often use a hash table to mark seen URLs. Then if we encounter a website we have already downloaded, we can just skip it.

Computationally, a hash table brings several conveniences to our cause. It requires $O(1)$, or constant time to determine whether a URL already exists in the table. We can also insert unseen URLs into the table in constant time. On a single download server, a hash table is very easy to implement. In practice, however, coordinating a unified table across thousands of servers poses a completely new level of difficulty. First, the hash table does not fit on a single server, so it must be stored in a distributed manner. Second, each download server will need to maintain this table before and after downloads, so the hash table's servers are the bottleneck of the whole system. I frequently mention this latter challenge to Google applicants and elicit solutions for alleviating the bottleneck.

This task has various solutions, none absolutely correct, but each to varying degrees of feasibility. Strong answers generally mention two points. First, servers should be assigned non overlapping sectors of URLs to download, so downloading does not require consensus with other servers. A clear division of labor may involve assigning a specific range of URLs to each server. Second, all queries to the hash table should be batched, to minimize communication costs. In these ways, communication is vastly reduced, as is the load on the hash table servers.

9.4   Summary

For many years after graphs were first invented, their applications in the real world never exceeded a thousand vertices or so (e.g., road maps, railroad tracks). At that time, graph traversal was a simple problem, so few researchers in industry studied graphs. Even few computer science students chose this track of research, because they might never encounter graphs again. With the arrival of the Internet, however, graph traversal suddenly became the backbone of web search. Likewise, many mathematical tools have emerged to prominence out of complete obscurity. Thus, many still dedicate their lives to esoteric mathematics, because it is beautiful, and there will eventually be a place for it.

Chapter 10

PageRank: Google's democratic ranking technology

Today, most search engines return thousands of results for each query, but which of the thousands would users like to see the most? A search engine's quality is determined by how well it answers this question. Here, we will spend two chapters explicating this challenge. In general, search result rankings depend on two criteria: page quality and content relevance. This chapter introduces methods of ensuring page quality, while the next will focus on relevance to the search keywords.

10.1   The PageRank algorithm

You may have heard that Google's original claim to fame was an algorithm called "PageRank." This technology allowed a revolutionary leap in search quality around 1998, and it more or less solved the problem of web search. Afterwards, Google's search engine was widely regarded as the best, and the rest is history. Since then, PageRank has been held in reverence.

PageRank was by no means the first attempt to sort search results. The first company that recognized the importance of sorting was not Google, but Yahoo. Yahoo's cofounders, Jerry Yang and David Filo, used a directory structure to organize information on the Internet (for more of this history, I refer readers to my other book, *On Top of Tides*). Due to limitations in computer speed and capacity of that time, Yahoo and similar early search engines faced a common problem: their content was too few, and the search terms too vague. It was difficult to find relevant information because only the most common search terms were indexed. Sometime before 1999, I remember having to confer with several search engines for literature, since no search engine produced more than a few results.

Later, DEC built the AltaVista search engine, which contained more web pages than any other search engine. Using just a single Alpha server, AltaVista had indexed all the words on every web page. While the quantity was much improved, however, the quality of search results was still low. Most results were not relevant to the query, and useful information was often hidden behind pages and pages of bogus results. AltaVista solved the issue of coverage, but failed to rank their search results effectively.

At that point, AltaVista and its contemporary, Inktomi, both realized that the quality of web search also depended on the sorting of search results. They attempted several implementations of search ranking, but none of their methods were mathematically well-grounded. Those approaches generally applied techniques based on anchor text (text that appears as a link), which was a common method at the time. In 1996 my colleague, Scott Weiss, (later a professor at the College of William and Mary) used the number of links as a search ranking factor in his doctoral dissertation.

We mention the methods above for historical completeness, but their contributions towards improving search quality were feeble. Instead, the turning point of search came from Google's cofounders, Larry Page and Sergey Brin. So we return to their legendary PageRank algorithm—how does it work? Put simply, PageRank is a democratic voting mechanism. For example, we might have 100 people in a room who all claim that they are the next President of the United States. So who will be the real President? There may be several top candidates, but we can ask the public to vote. If everyone agrees that Donald Trump will be the next President, then we would rank Trump as the top candidate. Likewise, we can vouch for a web page's quality by linking to it. If a single web page is universally acknowledged and trusted, then many pages link to it, and its ranking will be high. This idea is the essence of PageRank.

In practice, Google's PageRank is slightly more complex. For example, a tweet by Ivanka Trump carries more influence than one by your neighbor's teenage child. To distinguish between these different links, websites with higher rankings give more weight to their outgoing links than those with lower rankings. This strategy is similar to the real world, where shareholders with different voting powers influence the final vote to varying degrees. The PageRank algorithm takes this idea into consideration, where voting power is determined by the rank of the incoming page.

To give a concrete example, a page's rank should be determined by the sum of incoming weights $X_1, X_2, \ldots, X_k$. In Figure 10.1, the rank of page Y is

$$\text{pageRank} = 0.001 + 0.01 + 0.02 + 0.05 = 0.081.$$

Although Page and Brin do not disclose which of the two discovered the PageRank algorithm, as far as I know, Page came up with this weighted ranking idea. According to Page, the incoming weights should be equal to the ranks of the pages themselves, so we come to a circular problem: what came first, the chicken or the egg (or weights and ranks)?

It was likely Brin who solved this problem. He converted the graph problem into a two-dimensional matrix multiplication problem and provided an iterative solution. First, assume that all pages have equal weights. Update based on these initial weights and repeat. Together, Page and Brin proved that this algorithm theoretically converges to the true rank, independent of initialization. Also worth mentioning is that this algorithm does not require human supervision; it automatically runs until convergence.

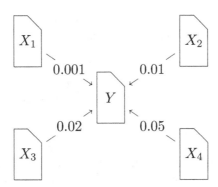

FIGURE 10.1: PageRank algorithm example.

Now that the theoretical problems were out of the way, practical problems emerged. There is an enormous number of pages on the Internet, so the aforementioned two-dimensional matrix grows quadratically with that huge number. If there are a billion web pages, there are 10^{18}, or one quintillion entries. We either require a very big computer to crunch those numbers, or as Page and Brin implemented, sparse matrix optimizations. When only the non-zero entries were stored in memory, Page and Brin were able to realize their page ranking algorithm with much less computation.

As the Internet grew, PageRank became computationally intensive to run. Early versions of the algorithm ran with half-manual, half-automatic parallelism across machines, so it took a long time to run through each iteration of Page-Rank. In 2003, Google engineers, Jeffrey Dean and Sanjay Ghemawat, solved the parallelism problem with MapReduce, a distributed computation framework. Parallel computation was automated, calculations were sped up, and update cycles were vastly shortened.

When I joined Google, Page chatted with us new employees about the early days of PageRank with Brin. He said, "at that time, we thought the Internet was a huge graph, where every website was a node and every link, an edge. If the Internet could be described as a graph or matrix, then maybe I could write a Ph.D. thesis from this idea." Thus blossomed the PageRank algorithm and so much more than a thesis. At the end, we jokingly asked him, why "PageRank," not "BrinRank"?

PageRank stood out as an algorithm because it treated the Internet as a whole network, rather than a collection of websites. This fact inadvertently considered the Internet as a system, instead of a set of moving parts. In contrast, previous attempts at search focused on individual webpages. The prevailing thought of that time was that each page contained enough information about itself to judge whether it was relevant to any search query. Although some of Page's and Brin's contemporaries considered using the links between pages as a metric of search quality, no one else created more than a patchwork solution. Only Page and Brin tackled search from the network level.

After it was released, PageRank had an astronomical impact on search results. From 1997-1998, you would be lucky to find two or three relevant web pages among ten search results. In contrast, Google, then still a fledgling at Stanford University, attained seven to eight quality results out of every ten. This difference in quality is akin to offering a flip phone user the latest iPhone. Thanks to PageRank, Google was easily able to outcompete all existing search engines.

Today, any commercial search engine can produce seven to eight relevant results out of ten, so search relevance is no longer a barrier to entry. Instead, the most important data is to collect are user clicks, which determine search engine quality today. As a result, new search engines have little accuracy improvements to offer, that would be noticeable enough to consumers. This is why companies like Microsoft and Yahoo struggle to maintain a market share in search.

10.2 Extended reading: PageRank calculations

Suggested background: linear algebra

Let vector

$$B = (b_1,\ b_2, \ldots,\ b_N)^T \tag{10.1}$$

be the rankings for pages $1, 2, \ldots, N$, where N is the total number of pages. Let matrix

$$A = \begin{pmatrix} a_{11} & \cdots & a_{1n} & \cdots & a_{1N} \\ \vdots & & & & \vdots \\ a_{m1} & \cdots & a_{mn} & \cdots & a_{mM} \\ \vdots & & & & \vdots \\ a_{M1} & \cdots & a_{Mn} & \cdots & a_{MM} \end{pmatrix} \tag{10.2}$$

contain the number of links between web pages, where a_{mn} is the number of links from page m to page n.

Suppose A is a known value and B is an unknown we are trying to calculate. If B_i is our estimation of B after the ith iteration, then

$$B_i = A \cdot B_{i-1} \tag{10.3}$$

where we initialize B_0 as a uniform distribution,

$$B_0 = \left(\frac{1}{N},\ \frac{1}{N}, \ldots,\ \frac{1}{N} \right)^T.$$

Obviously, we can use Equation 10.3 to find $B_1, B_2, \ldots$ (computationally expensive but direct). We can prove that B_i eventually converges to the true B, for which $B = B \times A$. When the difference between B_i and B_{i-1} is sufficiently small, we stop computing. Convergence generally takes around 10 iterations for PageRank.

Other elements to consider include the Internet's sparsity. To avoid zero probabilities, we assign a small constant α to smooth out values which would otherwise become zero. Thus, Equation 10.3 becomes

$$B_i = \left(\frac{\alpha}{N} \cdot I + (1 - \alpha)A\right) \cdot B_{i-1} \tag{10.4}$$

where N is the number of web pages, α is a small positive constant, and I is the identity matrix.

PageRank's computation primarily involves matrix multiplication, which can be broken up into many small tasks on separate machines. Chapter 29 discusses the specific methods for distributed computation using MapReduce.

10.3   Summary

Today, Google's search engine is a lot more complex than the original, but PageRank remains central to all of Google's algorithms. In academia, PageRank is recognized as one of the most important contributions to information retrieval, and it is even included in the curricula of many universities. For this algorithm, Page was elected to the National Academy of Engineering at the age of 30, adding another college dropout to the list (the first was Bill Gates). In addition, the PageRank patent led to two practical implications. First, other companies could not use the technology without infringing on Google's intellectual property. Google was a young company then, and it needed all the protection it could muster. Second, it gave Stanford University over 1% of Google's shares, which amounts to over 1 billion US dollars today.

Bibliography

1. Brin, Sergey and Page, Lawrence. The Anatomy of a Large-Scale Hypertextual Web Search Engine, http://infolab.stanford.edu/~backrub/google.html.

Chapter 11

Relevance in web search

So far, our discourse on search engines has covered how to automatically download web pages (web crawlers) and how to measure the quality of web pages (PageRank). Now we will add the final ingredient—how to select relevant web pages to a search query. With these three pieces of knowledge, anyone with coding experience can build a simple search engine, for a personal or community project.

When I wrote this article in 2007 for Google's blog, algorithms and technology were still more important than data, so the quality of search engines depended on their engineering. Today the scene is somewhat different. Commercial search engines are technologically mature, and they have accumulated extensive user data. Probabilistic models based on click data contribute the most to today's search results. In addition to click data, other aspects to search fall into four categories:

1. A large, complete index. If you do not index a web page, then the world's best algorithm cannot help you.

2. Page quality metrics, such as PageRank. Compared to ten years ago, PageRank holds a much smaller role in quality metrics. Today, a variety of factors contribute to a web page's quality. For example, click through rate may score high on PageRank, but their contents are less authoritative than their ranks would suggest.

3. User preferences. Different users naturally have different preferences for the content they see. A good search engine will customize search rankings to each user's taste.

4. Relevance metrics. These are the focus of this chapter.

Suppose we were interested in modern classical music, so we wanted to research Stravinsky's *Rite of Spring*. Our first step is to discover all pages containing the three words "rite," "of," and "spring." Any decent search engine can provide hundreds of thousands, or even millions, of results. As of 2018, Google returned 1,650,000 results for that search query. Unfortunately for search engines, any sane user will not look past the first few pages. We must ensure that the best results are placed on the front page. Not only should these results be high quality pages on their own, but they should also contain content relevant to "rite of spring." Chapter 10 already covered the quality metrics for web pages, so here we provide a tool that measures closeness between search queries and web results.

11.1   TF-IDF

The phrase "rite of spring" can be split into three words: "rite," "of," and "spring." Intuitively, web pages that contain more occurrences of these words are more relevant than web pages without them. Of course, a caveat is that longer pages are more likely to contain keywords, so we must account for document length. Thus, we divide the number of occurrences by the length of the page, to obtain term frequency. For example, suppose a website has 1000 words, of which "rite" appears 2 times, "of" appears 35 times, and "spring" appears 5 times. Correspondingly, we have frequencies 0.002, 0.035, and 0.005. Summing those numbers up, we obtain the term frequency of this document, 0.042, for the query "rite of spring."

Term frequency presents an easy way to measure the relevance between web pages and search queries. Formally, if we have N words $w_1, w_2, \ldots, w_N$ and corresponding term frequencies $TF_1, TF_2, \ldots, TF_N$, then the term frequency (TF) of that page for the query is

$$TF_1 + TF_2 + \quad + TF_N. \tag{11.1}$$

Careful readers may have noticed another caveat in this equation. In the above example, the word "of" contributed over 80% of the total term frequency, and "of" is quite useless a word to find. Common words that do not add to a document's subject are known as stop words. In English, other common stop words are "the," "is," "and," and "a," etc. Stop words are often ignored in text analysis because they detract from real insight. If we ignore stop words for the above example, then the TF becomes $0.002 + 0.005 = 0.007$.

Finally, we patch up another shortfall of our model. In modern English, "spring" is relatively common, while "rite" is rare and archaic. (When did you even last say "rite" to someone?) Therefore in search results, appearances of the word "rite" should carry more weight than those of "spring." This weight must satisfy two conditions:

1. Words that give more information about a topic should carry more weight. On the contrary, words that are less discriminatory should carry less weight. For example, "rite" signals more about a web page's content than "spring," so "rite" is given a higher weight.

2. Stop words should contribute zero weight.

If a word appears in very few web pages on the Internet, then that word can drastically narrow down the search space, so it should be weighed more.

Conversely, if a word appears everywhere, then we still do not know much after seeing that word, which should be weighed less.

Formally, suppose word w appears in D_w pages across the Internet. The larger D_w is, the smaller w's weight should be, and vice versa. In information processing, the most common metric used to determine w's weight is inverse document frequency (IDF), computed as $log(D/D_w)$, where D is the total number of pages. Suppose the Internet contains 1 billion pages, and the word "the" appears in all 1 billion. Then its IDF is

$$log\left(10^9/10^9\right) = log\left(1\right) = 0.$$

On the other hand, the "rite" may only appear in 2 million pages, so its IDF is

$$log\left(\frac{10^9}{2 \ 10^6}\right) = 8.96.$$

Finally, the word "spring" is more common, and it appears in 500 million pages (an assumption of course). So its IDF would be

$$log\left(\frac{10^9}{5 \ 10^8}\right) = log\,2 = 1.$$

In other words, finding "rite" in a document gives you 9 times more information than "spring." Factoring in IDF, Equation 11.1 becomes

$$TF_1 \ IDF_1 + TF_2 \ IDF_2 + \quad + TF_N \ IDF_N. \tag{11.2}$$

Our sample webpage has a correlation of 0.0161 with "rite of spring," where "rite" contributes 0.0126 and "spring" contributes 0.0035, consistent with our intuition.

The concept of TF-IDF is widely recognized as the most important invention in information retrieval, with widespread applications in document classification and other fields. Though famous today, TF-IDF had a rough start. In 1972, Karen Spärck Jones* from the University of Cambridge published "A statistical interpretation of term specificity and its application in retrieval," in which TF-IDF was first proposed. Unfortunately, she did not thoroughly explain its theoretical underpinnings, including why IDF should be an inverse logarithmic function, rather than a square root, for instance. She also did not investigate further on this topic, so later researchers generally do not cite her essay when referring to TF-IDF (and many do not know of her contribution).

In the same year, Stephen Robertson from the University of Cambridge wrote a two-page explanation of Spärck Jones's work, but its content was abysmally confusing. Finally, it was Cornell University's Gerard Salton who brought

---

*Spärck Jones is best known for her quote, "Computing is too important to be left to men."

TF-IDF to the public eye. He wrote several papers and a book discussing the matter. Augmented by Salton's own fame, that book is most commonly cited as the source of TF-IDF. In fact, the most important award in information retrieval is named for Salton, and many attribute TF-IDF's widespread usage to him.

Fortunately, the world has not forgotten Spärck Jones's contribution either. In 2004, to commemorate the 60th anniversary of the Journal of Documentation, the article was reprinted in full. Here again, Robertson wrote an article explaining IDF through Shannon's theory of information. His article was theoretically correct, but poorly written and lengthy (18 pages), complicating the whole matter. In fact, most scholars of information theory had already discovered that IDF is the Kullback-Leibler divergence of the probability distribution of keywords under certain assumptions (see Chapter 6). As such, the measure of relevance in information retrieval ties back to information theory.

Nowadays, search engines have added subtle optimizations to TF-IDF to improve its performance. For those interested in writing a search engine for fun, however, TF-IDF is sufficient as is. Given a query, the overall ranking of a web page is roughly determined by the product of its TF-IDF and PageRank.

11.2   Extended reading: TF-IDF and information theory

Suggested background knowledge: information theory and probability theory.

In a given query, the weight of each keyword should reflect the amount of information that keyword provides. A simple method is to use the keyword's entropy as that weight. That is,

$$I(w) = -P(w) \log P(w)$$

$$= -\frac{TF(w)}{N} \log \frac{TF(w)}{N} = \frac{TF(w)}{N} \log \frac{N}{TF(w)} \qquad (11.3)$$

where N is the size of the corpus. Since N is a constant, we can omit N and simplify the equation to

$$I(w) = TF(w) \log \frac{N}{TF(w)}. \qquad (11.4)$$

However, Equation 11.4 has a flaw. Suppose two words appear the same number of times, but the first primarily appears in one article, while the latter is dispersed across many articles. Obviously, the first word is more discriminative about its article, so it should carry more weight. A better formula for weights should thus respect the discriminability of keywords. To give such a formula, we make the following assumptions.

1. The size of each document M is approximately constant. That is,

$$M = \frac{N}{D} = \frac{\sum_w TF(w)}{D}.$$

2. If a keyword appears in some document, then its frequency in that document is irrelevant. A word either appears $c(w)$ times or not at all. Note that $c(w) < M$. We can then simplify equation as

$$TF(w) log \frac{N}{TF(w)} = TF(w) log \frac{MD}{c(w)D(w)}$$

$$= TF(w) log \left(\frac{D}{D(w)} \frac{M}{c(w)} \right). \qquad (11.5)$$

The difference between the amount of information received and TF-IDF is the right side of Equation 11.6. Since $c(w) < M$, the right-hand term is a positive, but decreasing function of $c(w)$. We can now rewrite this difference as

$$TF\text{-}IDF(w) = I(w) - TF(w) log \frac{M}{c(w)}. \qquad (11.6)$$

The more information w brings, the larger $I(w)$, so the larger its TF-IDF value. Likewise, the higher w's average occurrence count, the smaller the $log\ M/c(w)$ term, so the larger its TF-IDF, in accordance with information theory.

11.3   Summary

TF-IDF is a measure of search relevance with strong theoretical bases. Therefore, even those new to search can directly use TF-IDF for satisfactory results. Though many search engines fine-tune TF-IDF to their own applications, they are all rooted in the same theoretical principles.

Bibliography

1. Spärck Jones, Karen. "A statistical interpretation of term specificity and its application in retrieval". Journal of Documentation 28 (1): 11–21, 1972.

2. Salton, G. and M. J. McGill, Introduction to modern information retrieval. McGraw-Hill, 1986.

3. Wu, H.C., Luk, R.W.P., Wong, K.F., Kwok, K.L. "Interpreting tf-idf term weights as making relevance decisions". ACM Transactions on Information Systems 26 (3): 1–37, 2008.

Chapter 12

Finite state machines and dynamic programming: Navigation in Google Maps

In 2007, when I first published the Google blog articles, location services were neither ubiquitous nor reliable. Smartphones were not as common as they are now, so digital maps were not integrated with users' locations. Map search only contributed to a fraction of web search as a whole.

Today, in contrast, location search is commonplace in everyday life—from searching for directions, to discovering nearby restaurants and sharing your location on social media. Map search (confirming locations, finding directions, etc.) remains the foundation of these services. To reduce the number of chapters in this book, I combine the topics of finite state machines and dynamic programming into an overview of map search.

On November 23, 2008, Google, T-Mobile, and HTC announced G1, the first 3G, Android-based smartphone. Compared to Apple's iPhone, released the previous year, the G1 could not hope to compete in appearance or user experience, despite their similar prices. Nonetheless, that phone attracted many users. Its killing feature? An integrated global positioning service (GPS), which provided global navigation. Satellite navigation had already existed for vehicles by the turn of the century, but it came with a high price tag. In 2004, a Magellan portable navigation system cost $1000 (available for $100 to $200 today). Earlier smartphones also advertised GPS features, but their functionality was not up to par. The Android phone filled in this market vacuum, with a GPS system comparable to satellite navigation technologies and address recognition technology that was much better than existing in-car GPS products. Using finite state machines, the Android address recognition system tolerated input errors and deviations, whereas GPS products required exact addresses. As a result, Magellan's stock plunged by 40% the day the G1 was released.

We can simplify a smartphone's navigation system into three components: satellite positioning, address verification, and route planning. Global satellite positioning uses traditional techniques, so we do not discuss them here. Address identification is covered in the first half of this chapter, alongside finite state machines. Finally, route planning (e.g., fastest path from starting point to destination) is introduced in the latter half of the chapter, with dynamic programming.

12.1   Address analysis and finite state machines

Address recognition and analysis are essential to map search services. Judging the validity of an address, while extracting geographic information (city, state, etc.), may seem trivial tasks for a human, but they are quite cumbersome for computers. For example, when I worked at Tencent in Shenzhen, China, I received mail addressed in these various ways:

Tencent Headquarter Building, Shenzhen City, Guangdong Province

Tencent Building, Nanshan District, Shenzhen Science and Technology Park, Guangdong Province, 518057

Tencent Corporation, 518057 Shenzhen Science and Technology Park, Nanshan District

Tencent, 518000 [probably did not know the exact number] Shenzhen Science and Technology Park, Nanshan District in Guangdong Province

These addresses are all a bit different, but I received all these packages, so the postman obviously recognized them. However, a programmer would have a harder job, writing a parser that supports all of these formats. Addresses, though simple for natural language, are still context-dependent. For example, consider these two examples:

1. 190 Massachusetts Ave, Cambridge, MA

2. 190 Massachusetts Ave, Boston, MA

When a parser sees "190 Massachusetts Ave," it must read on to discern which city the address refers to. After all, one address refers to a scrumptious bakery, while the other points to a costume and wig shop. As we have previously mentioned, context-sensitive grammars are complicated and tedious. Its parsers are time consuming and inelegant—and even so, they cannot cover all corner cases. Suppose our parser recognizes the number-street-city model. Fragile as it is, the parser would fail on the following input:

100 meters northeast of the intersection of Massachusetts Avenue and Memorial Drive.

Unfortunately for software engineers, many businesses actually advertise their addresses as such. Luckily, addresses constitute a relatively simple grammar, compared to most of natural language, so there are several ways to dissect them. The most effective of these is the finite state machine.

A finite state machine (FSM) is a special type of directed graph (see Chapter 9). The graph contains states (vertices) and directed transitions (edges) from

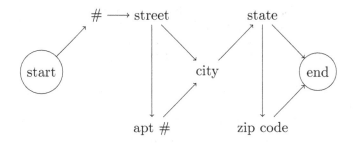

FIGURE 12.1: Finite state machine that recognizes addresses.

state to state. Figure 12.1 depicts an example of a finite state machine that recognizes American addresses.

Each FSM has a start state, a terminal state, and a number of intermediate states. Each edge presents a condition for entering the next state. In the above example, suppose we are at the state "street." If the next word contains a number, then we move to the "apartment number" state. Otherwise, if the next word describes a city, then we move to "city." If we can travel from the start state to the terminal state through a series of legal transitions, then we have a valid address. According to this FSM, "77 Massachusetts Ave, Cambridge, MA" is accepted, while "77 Massachusetts Ave, MA, Cambridge" is rejected (since we cannot return to "city" from "state").

We must solve two problems before we can effectively leverage FSMs for address recognition: given a set of addresses, construct an FSM that recognizes them; and given an FSM, provide an address matching algorithm. Circular as they are, these problems are solvable using existing algorithms, so we need not worry about them.

With such a model at hand, we can provide location services to the Internet. From web pages, we extract relevant address data and create a database of discovered places. Afterwards, a user can input a search query with an address, and the search engine will pick out the location details, as well as related keywords. For example, if you type in "restaurants in Boston, MA," Google will automatically identify "Boston, MA" as your location and "restaurants" as your object of search.

The FSM in Figure 12.1 has several drawbacks in practice. If the user inputs a non-standard address format or makes a typo, the FSM fails, just as a strict set of grammar rules. In computer science, the earliest applications of FSMs could avoid this issue; compilers parsed programming languages, and runnable code is always syntactically correct. When applied to natural language, however, FSMs must be modified to tolerate human error. We want FSMs to "fuzzily" match addresses and give a probability of correctness. To this end, scientists developed probabilistic FSMs, which are equivalent to the discrete Markov chains in Chapter 6.

Before the 1980s, many used probabilistic FSMs in their research, but most were designed for a single application only. As probabilistic FSMs became widespread in natural language processing, many scientists developed generic libraries for FSMs. Of those, the most successful library was written by former AT&T scientists, Mehryar Mohri, Fernando Pereira, and Michael Riley. After many years of hard work, the three compiled an outstanding library for probabilistic FSMs in C.

Following in the AT&T tradition of providing code for academic research, all three decided to share that library with their peers. Soon after, however, AT&T's fortunes declined, and the three parted ways from AT&T. Mohri took up a post at New York University; Pereira became the chair of computer science at the University of Pennsylvania, and later research director at Google; and Riley directly joined Google as a research scientist. During this transitional period, AT&T stopped releasing their C library for free. Though the three had published papers regarding the library, they omitted many implementation details. After the FSM library was removed, academics recreated its functionality from literature, but none perfectly. For many years, it was difficult to replicate the performance of that original library. In the recent decade, open source software has become quite influential, so AT&T once again released this library to the public.

Finite state machines are easy to understand but nontrivial to implement. They require thorough knowledge of the theory and the engineering, as well as strong programming ability. Therefore, I recommend that interested readers use the open source FSM code, rather than implementing a library from scratch.

Beyond address matching, FSMs may actually be used towards a variety of applications. A typical example would be Google Now, Google's former personal assistant software on smartphones. At the core of Google Now was a finite state machine that considered a user's personal information, location, calendar, and voice or text inputs. With a combination of these data, Google Now could discern a user's desires and provide the relevant information and/or services (e.g., open the phone app, set a reminder, search for yummy food). Inside Google Now, the underlying engine was functionally equivalent to AT&T's probabilistic FSM.

12.2   Global navigation and dynamic programming

The key behind global navigation is the application of dynamic programming towards graph search algorithms.

Recall that a graph is a collection of vertices and edges, in which edges connect vertices. If each edge is assigned a weight, then the graph is known as a "weighted graph." For example, the US highway system in Figure 12.2 is such a graph where edge weights correspond to the distance between cities (in miles). Other reasonable edge weights for that map include travel time, total cost of tolls, or miles of highway.

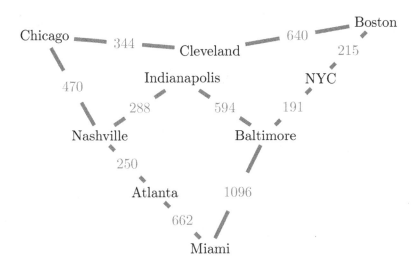

FIGURE 12.2: The US highway system as a weighted graph, where edge weights represent distance in miles.

A common problem in graph theory is the shortest path problem: given a start and target vertex, find the path that minimizes the total weights of included edges. In our US highway example, we might want to find the most direct route from Boston to New York City. Obviously, we would rather take the road straight to New York City, instead of traveling around the country, down to Miami, and back north. Naively, we could find this optimal path by enumerating all paths, then selecting the shortest one. Unfortunately, this method only works for graphs where the number of vertices is countable in single digits. After including several dozen vertices, even the most powerful computers cannot reasonably list out all the paths. In fact, the asymptotic complexity of this method grows exponentially as the number of vertices. For each additional city added, we (at least) double the time spent on listing paths. Since commercial navigation systems find the desired route within seconds, they obviously do not enumerate paths behind the scenes.

Instead, your GPS uses dynamic programming (DP).* The gist of dynamic programming is simple: we can break up a difficult problem into easier subproblems, solve those subproblems, and combine them into the original answer. Important to note, the subproblems are smaller versions of the original, of approximately the same form.

Let us consider the US highways example again. Suppose we wanted to find the shortest path between Boston and Atlanta, and we currently believe that

---

*A small note on terminology—in dynamic programming, "programming" refers to "planning," rather than "coding." It was first developed for resource allocation in the military.

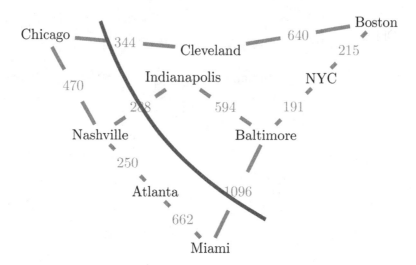

FIGURE 12.3: All paths from Boston to Atlanta must cross the dark line.

path A is the shortest. If path A contains Baltimore, then the component of A from Boston to Baltimore must be the shortest path between the two cities. Otherwise, if there is a shorter way to reach Baltimore, then some path B contains this shortcut. As a result, B must be shorter than A, which would not actually be the shortest. Therefore, to prove that a path is shortest, it is either the only path, or all other paths are longer. Breaking down the shortest path problem, if any route is the shortest, then its constituent components must also be the shortest between their cities.

Concretely, if we want to know the shortest path from Boston to Atlanta, we must first find the shortest path from Boston to Baltimore. Of course, before we know the shortest path itself, we cannot know that it passes through Baltimore in the first place. To find the proper intermediate cities, we can draw a line across the map, separating Atlanta from Boston, as in Figure 12.3.

All paths from Boston to Atlanta must cross the line at some point, so the shortest path contains one of the cities before the line (Baltimore, Indianapolis, or Cleveland). If we find the shortest path from Boston to these three cities, then we have found a portion of the shortest path to Atlanta. In this way, we can break up the shortest path problem into smaller subproblems. We keep finding shortest subpaths by drawing a line between source and destination, until we have found all the shortest paths we care about. Using dynamic programming, there are at most $O(n^2)$ subproblems, one for each pair of cities. For comparison, the naive method (of listing all paths) would have encountered approximately three new cities per "line." Thus, the number of paths would have grown as $O(3^n)$, which explodes as the number of cities grows large.

As seen here, the correct model can vastly reduce the complexity of hard problems. Such is the magic of mathematics.

12.3   Finite state transducer

Suggested background knowledge: graph theory

Beyond address recognition, finite state machines have applications, including speech recognition decoders, compilers and circuits. Therefore, we provide their formal definition here.

Definition 1. *A finite state machine is a 5-tuple (Σ, S, s_o, δ, f), where*

- *Σ is the set of input symbols,*

- *Q is a non-empty, finite set of states,*

- *q_0 is the special starting state,*

- *δ is the transition function that maps $\delta : Q \times \Sigma \to Q$, and*

- *f is the special terminal state.*

Here, δ need not be defined for all pairs $(q, s) \in \Sigma \times S$. That is to say, there are some input strings that may not be accepted by a given finite state machine. For example, in Figure 12.1 (address recognizer), we may not give an apartment number after providing the state. Should we attempt to, the finite state machine will reject our address. If an input string can start at state q_0, travel through some intermediate states, and end up at the terminal state f, then it is accepted by the finite state machine. If the input ever reaches a state with no valid transitions, then it is rejected.

Finite state machines play an important role in speech recognition and natural language processing. These fields use a special model, known as the weighted finite state transducer (WFST), described here. In a finite state transducer, each state is described by its input and output symbols, as shown in Figure 12.5.

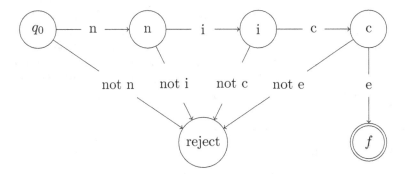

FIGURE 12.4: If we ever reach a state with no valid transitions, we can direct inputs to a "rejecting" state. This finite state machine accepts the word "nice."

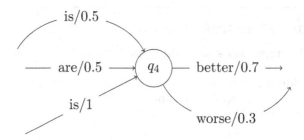

FIGURE 12.5: Single state in finite state transducer.

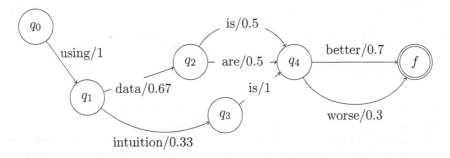

FIGURE 12.6: This WFST recognizes sentences that comment on the utility of intuition or data.

In Figure 12.5, state q_4 is defined as "inputs of is/are, outputs of better/worse." Regardless of the rest of the sequence, as long as the symbols before and after match, then an input can enter this state. Other states can have different inputs and outputs. Furthermore, if transition probabilities differ among the states, then we have a weighted finite state transducer. With respect to natural language processing, you may recall that bigram model from Chapter 2 also considers pairs of words. Thus, the WTST naturally maps itself to a sentence. Any sentence recognizable by a speech recognition system can be represented by a WFST. Each path through the WFST is a potential sentence, and the sentence with the highest probability is the speech recognition result. To find that sentence, we use dynamic programming, as described earlier.

12.4   Summary

Finite state machines and dynamic programming are useful beyond location services. They contribute to speech recognition, spelling and grammar correction, controls, and even biological sequence analysis. In later chapters, we will also introduce their application to pinyin input, the standard method of digitally inputing Chinese.

Bibliography

1. Mohri, Mehryar., Pereira, Fernando., and Riley, Michael. *Weighted finite-state transducers in speech recognition*, Computer Speech and Language, V16-1, 69–88, 2002.

Chapter 13

Google's AK-47 designer, Dr. Amit Singhal

Firearms enthusiasts or those who have seen Nicholas Cage's *Lord of War* may resonate with that movie's introduction:

> [The AK-47 is] the world's most popular assault rifle, a weapon all fighters love. An elegantly simple nine pound amalgamation of forged steel and plywood. It doesn't break, jam or overheat, it will shoot whether it's covered in mud or filled with sand. It's so easy, even a child can use it, and they do.—*Lord of War*

In the field of computer science, algorithms that withstand the test of time are like the AK-47: effective, reliable, uncomplicated, and easy to implement or operate. Ex-Google Fellow, Amit Singhal, was the designer of Google's AK-47 algorithms. The "A" in Ascorer, Google's internal search ranking system, was named after him.

On my first day at Google, I started my four years of pleasant cooperation with Amit Singhal, and he has been my mentor ever since. Singhal, Matt Cutts, Martin Kaszkiel, and I worked together to solve the problem of search engine spam.* We realized that the majority of search engine spam had commercial motives, which were understandable. Thus, we needed to build a classifier that detected commercial intents. Coming from an academic background, I always strove for the perfect solution. I could easily build a classifier that would be usable and elegant, but would take three or four months to train. At that time, Google had no parallel computing frameworks in place (i.e., MapReduce did not exist), so complicated machine learning was costly to train.

Singhal declared that a straightforward solution was enough. He asked me how long it would take to implement one of the simplest classifiers, and I answered, one weekend.† I delivered the classifier on Monday and asked if he wanted me to perfect the model. He looked at the results and answered, "Good enough, good enough. The best solutions are always simple." We had made a bet with Wayne Rosing, vice president of engineering: if we could reduce spam by 40%, then he would bestow upon us an award of engineering and send

---

*We will delve into the technicalities of anti-spam in later chapters.

†When Google was a startup, we typically did not take the weekends off.

our four families to Hawaii for a five-day vacation. Following the principle of simplicity, we reduced the amount of spam by half within the month. Rosing delivered.

Sergey Brin asked me how I crafted this classifier in such a short amount of time. I told him that we built from basic principles, and he replied, "oh, it's like the AK-47 automatic rifle" simple, efficient, and effective. Our classifier was designed to be compact and fast—it used minimal memory and could be run on just a few servers. Even after I left Google, the design still provided value to the company. Incidentally, this classifier was also the first patent Google obtained in anti-spam.

Later, Singhal and I completed many projects together, including the design and implementation of a better ranking system for Chinese, Japanese, and Korean (CJK). In 2002, while Google supported search in approximately 70 languages, they all relied on the same ranking algorithm. Efforts in internationalization did not extend past translating foreign documents and encoding them with the appropriate character set. At that time, Singhal approached me to design a brand new search algorithm for Chinese, Japanese, and Korean. To be quite honest, I was not interested in language-specific search, but back then, I was the only Chinese person studying natural language processing at Google. Furthermore, CJK search results were awful compared to those in English, so this task fell upon me.

Taking inspiration from the previous anti-spam project, I opted for the most basic solution that would work well. After it was implemented, my method *did* attain excellent performance, but it used more memory than feasible. Google did not own as many servers as it does now, and it was unrealistic to purchase another fleet of machines solely for CJK search, which accounted for less than 10% of total traffic. Singhal proposed that we switch to a regression model, in place of the memory-intensive language model. The former would require no additional machines, but would only improve search quality by 80% of the latter. I was mildly unwilling to settle for this decreased performance, but Singhal explained to me that the additional 20% was icing on the cake. With a simpler model, we could ship the algorithm to Chinese customers at least two months earlier, and user experience would immediately improve. I accepted his suggestion. In early 2003, I released the first search algorithm specifically tailored towards CJK.

A year later, Google had purchased another fleet of servers, and I made further progress in model compression. At this time, I released my full algorithm to CJK users. One of Singhal's secrets for success in industry is his emphasis on the first 80%. If we can solve most of the customers' problems, then we should deliver the 80% now and fill in the 20% later. Many entrepreneurs fail not for lack of talent, but for lack of methodology. Those who pursue a grandiose, perfect product often fail to complete their work, which never sees the market.

Singhal always insisted on the simplest solution that would fit the bill. At Google, where talent was rampant, his philosophy met many opponents. Experienced engineers tended to underestimate the power of simple models. Between

2003 and 2004, Google recruited many natural language processing scientists from the world's leading labs and universities, including MITRE3, AT&T, and IBM. Many newcomers tried to "refine" Singhal's various "AK-47s" with the state of the art, but they soon realized that they could only make marginal improvements, at great computational cost. Outside of Google, researchers have also tried to build better search engines with more sophisticated methods (Microsoft, for instance), but they only complicate simple matters.

Perhaps Singhal's engineering philosophy is better summarized by Newton's famous quote:

> We are to admit no more causes of natural things than such as are both true and sufficient to explain their appearances.

That is, science should be simple. Likewise, Einstein wrote in *The Evolution of Physics*:

> In the whole history of science from Greek philosophy to modern physics there have been constant attempts to reduce the apparent complexity of natural phenomena to some simple fundamental ideas and relations. This is the underlying principle of all natural philosophy.

Einstein's principle also holds in the field of information. From a practical standpoint, Singhal's emphasis on straightforward methods allows the engineer to easily explain the rationale behind each step. Solid reasoning helps debug the system when issues arise and uncovers obvious areas for improvement.

Today, the quality of search has vastly improved since Page and Brin first started their research, nearly two decades ago. Large advances have given way to incremental progress. Subtle improvements in one field lead to worse performance in others. In this environment, we must be explicit about the tradeoffs and dig for the underlying explanations behind performance changes. With complicated methods, such as black-box machine learning models, we cannot pinpoint the cause of improvement or lack thereof. As a result, the long-term quality of search may not increase; it will be weighed down by unquantifiable tradeoffs. Thus, Singhal mandated that any means of improvement must be backed by solid reasoning. If a method cannot be explained, then it may cause unknown damage in the future.

Singhal's point of view differed from that of Microsoft and Yahoo, who treated search quality as a black box. Under his guidance, Google's search quality improved year after year. Of course, as Google obtained increasing amounts of data, it became more effective to adjust search parameters via machine learning methods. After 2011, Singhal emphasized search quality as a machine learning and big data problem. Continuing along with his methodology, however, Singhal remained adamant that models be explainable. Any model whose parameters and results were "black magic" was not allowed online.

It may seem that Singhal possessed extraordinary intuition or unnatural luck, but Singhal's success was owed to his extensive research experience, prior to Google. Singhal studied under Professor Gerard Salton for his Ph.D. and moved on to AT&T Labs after graduation. There, Singhal and two colleagues built a medium-sized search engine within six months. Although the number of pages indexed by Salton's search engine were incomparable to Google's, its search quality was still excellent. So at AT&T, Singhal thoroughly studied all the idiosyncrasies of search engines. His algorithms, though simple in appearance, were the fruits of long deliberations on real data. Even after he became a Google Fellow, Singhal continued to examine samples of poor search results every day. Constant firsthand experience with data proved invaluable. To young engineers, I encourage you to spend some time with your data every day. I find that many search engineers spend much less time examining poor search results than Singhal did, even after he came to fame.

Singhal encouraged young people to be bold and unafraid of failure. On one occasion, a recent college graduate panicked after she pushed buggy code onto Google's servers. Singhal consoled her by sharing his worst mistake at Google: he had once set the Internet's relevancy scores all to zero, so search was completely random for a short time. Many years down the road, that same young engineer had developed exemplary products for Google.

During his time at AT&T, Singhal had established himself among academics. Like many who have turned towards industry, however, Singhal was not content with writing papers for the rest of his life, so he left for Google, then only around a hundred employees. There, his talents shined; he rewrote Google's ranking algorithms, which have continually improved since. Reluctant to leave his two children at home, Singhal rarely attended conferences, but he was still widely regarded as the most authoritative figure in search quality. In 2005, Singhal was invited back as a distinguished alumnus to his alma mater, Cornell University, to celebrate the Department of Computer Science's 40th anniversary. Other recipients of this honor included the inventor of the RAID disk, Professor Randy Katz.

Singhal and I differ in many aspects of character and lifestyle—for one, he's a vegetarian and I'm not, so I never quite enjoyed his family's meals—but common to us both is the philosophy of simplicity.

In 2012, Singhal was elected to the Academy of Engineering and named the senior vice president of Google search. The same year, he recruited me back to Google with one request: start a project that would beat Microsoft for at least five years. Four years later, Singhal left Google after 16 years, to become an angel investor, a new odyssey in his career.

Chapter 14

Cosines and news classification

Some facts about the world are beyond our everyday imagination. The law of cosines and news classification may seem worlds apart, but the former drives the latter in modern technology.

In the summer of 2002, Google released its own "news" service. Unlike traditional media, Google News was neither written by reporters nor curated by editors. Rather, all the content was automatically discovered and generated. The key technology was automatic news classification.

14.1 Feature vectors for news

News classification, or more generally document classification, is the clustering of similar documents. If a human editor were given this task, he would read through all the documents, pick out the main ideas, and then group them by subject. Computers, on the other hand, do not read the news. Artificial intelligence experts or businessmen may tell you that computers can read, but in reality, computers can only perform arithmetic calculations, fast. In order for a computer to "compute" news, we must first convert documents into a set of numbers, then design an algorithm to find the similarity between these numbers.

Let us begin by finding a set of numbers (more specifically, a vector) that can describe news stories. News is the transmission of information, while words are the carriers of information—so the meaning of news is naturally linked to the semantics of its words. In the spirit of Leo Tolstoy's *Anna Karenina*: similar articles all share the same words, but each dissimilar article has its own differences.* Each news article has many words, not all equally important or discriminative, so how do we select the effective indicators? First of all, intuition tells us that rich, meaningful words must be more important than articles like "a" and "the," or conjunctions like "and" and "but." Next, we need to measure the importance of other words. Recall the concept of TF-IDF from Chapter 11 (search relevance). We can imagine that important words have high TF-IDF values, especially those related to an article's subject matter. Such words occur frequently within the article, so they have large TF-IDF values.

---

*The novel opens with: "Happy families are all alike; every unhappy family is unhappy in its own way."

TABLE 14.1: Index to word mapping.

Index	Word
1	a
2	an
3	as
...	...
456	dandelion
...	...
64000	zloty

Using TF-IDF, we can convert news into numbers. Suppose we have a dictionary of all words in the English language. We assign each word an index i. Now given some article, calculate the TF-IDF value of each word w_i and insert that value into position i of the document's vector. For example, if there are 64,000 words in our vocabulary,* then our dictionary might be similar to Table 14.1.

For a particular piece of news, the TF-IDF values for these words may be as depicted in Table 14.2

If a word does not appear in the article, then its corresponding value is 0. These 64,000 numbers form a 64,000-dimensional feature vector, each of which represents a single document. In each dimension, the magnitude of the vector represents the contribution of each word towards an article's meaning. When text is converted to this type of vector, a computer may quantify the similarities between articles, as shown in Figure 14.1.

TABLE 14.2: Index to word mapping.

Index	TF-IDF
1	0
2	0
3	0.0034
...	...
456	0.052
...	...
64000	0.075

---

*If we limit vocabulary size to 65,535, then we can represent each word with two bytes in the computer.

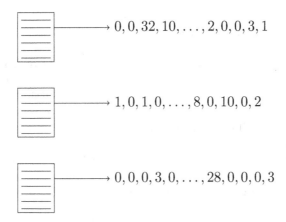

$$0, 0, 32, 10, \ldots, 2, 0, 0, 3, 1$$

$$1, 0, 1, 0, \ldots, 8, 0, 10, 0, 2$$

$$0, 0, 0, 3, 0, \ldots, 28, 0, 0, 0, 3$$

FIGURE 14.1: An article becomes a string of numbers.

14.2 Vector distance

No matter what language you grew up with, your language arts teacher probably taught you that specific ideas correspond to certain descriptive words. After thousands of years, mankind has developed the habit of writing with domain-specific vocabulary. As a result, each category of news uses some words more, others less. For example, financial news are likely to contain words such as: stocks, interest, bonds, banks, and prices. Reports on finance are less likely to contain: carbon dioxide, the universe, poetry, Nobel, and meat buns. If two articles share a common topic, then their vectors will be large in many of the same dimensions and small in others. Conversely, if two articles are unrelated, then there should be little overlap in their non-zero elements. In this way, we regard two articles as "similar" if their feature vectors have approximately the same shape.

Now we introduce a metric for quantifying this similarity. Those familiar with vector algebra know that a vector is simply a directed segment, from the origin to some point in multidimensional space.

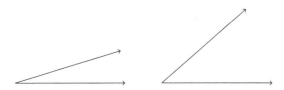

FIGURE 14.2: The angle between two vectors measures their "closeness." The vectors on the left are closer than the vectors on the right.

Different articles are generally not the same length, so their feature vectors will take on varying magnitudes. A text with 10,000 words simply has more *words* than one with 500, so the former will exhibit greater values in most dimensions. Thus, it makes little sense to compare the magnitudes of feature vectors. If we care little about a vector's magnitude, then we care a lot about its direction. When two vectors point in the same direction, they contain similar distributions of words. Therefore, we can use the angle between two vectors as a measure of their proximity. As illustrated in Figure 14.2, the left vectors form a smaller angle than the right vectors, so the left pair is closer.

To find the angle between two vectors, we apply the law of cosines, which you may recall from high school. This theorem states that given a triangle with edges a, b, c and opposite angles A, B, C:

$$\cos A = \frac{b^2 + c^2 - a^2}{2bc}. \tag{14.1}$$

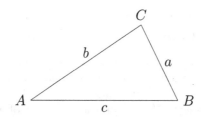

FIGURE 14.3: Law of cosines.

If we treat the two sides b and c as vectors rather than scalar quantities, then the above equation becomes

$$\cos A = \frac{\langle b, c \rangle}{|b| \cdot |c|}, \tag{14.2}$$

where the denominator is the product of the magnitudes and the numerator is the inner product. Expanding the terms, if we have two news articles with feature vectors

$$x = x_1, x_2, \ldots, x_{64000}$$
$$y = y_1, y_2, \ldots, y_{64000},$$

then the angle between x and y is

$$\cos \theta = \frac{x_1 y_1 + x_2 y_2 + \cdots + x_{64000} y_{64000}}{\sqrt{x_1^2 + x_2^2 + \cdots x_{64000}^2} \cdot \sqrt{y_1^2 + y_2^2 + \cdots y_{64000}^2}} \tag{14.3}$$

Since the feature vectors represent TF-IDF, they are non-negative, so the cosines take on values between 0 and 1. The corresponding angle ranges from 0 to 90 degrees. When the cosine of two vectors approaches 1, the two vectors are similar and likely fall into the same category. As the cosine decreases from 1, two feature vectors are farther apart, so the accompanying news articles are less unrelated. On the other extreme, when the cosine is 0, two vectors are orthogonal, so the news have nothing to do with each other.

Now that we have a method for converting news to vectors, and for calculating the similarity between vectors, we can begin our examination of news classification algorithms. Such algorithms are split into two types. Suppose we already know of some news categories, represented by their feature vectors $x_1, x_2, \ldots, x_k$. Given some unseen news article Y, we can select the closest cluster by cosine similarity. These categories can either be manually constructed (inaccurate and tedious) or automatically discovered (discussed later). Nonetheless, this first case is relatively easy to solve.

Instead, suppose that we only have news articles and no existing clusters. Our problem has become a lot harder. While I was at Johns Hopkins University, one of my peers, Radu Florian,* posed a bottom-up approach,[†] roughly depicted in Figure 14.4 and described below.

1. Compute all pairwise cosine similarities among the news articles. Merge all articles with a similarity score above some predetermined threshold into a subclass. This clusters N articles into N_1 clusters, where $N_1 < N$.

2. For each subclass, compute its representative feature vector (the vector mean). Replace original feature vectors with their subclass vector and repeat step 1, so that we have N_2 new subclasses, where $N_2 < N_1$.

We continue the process, as the number of classes decreases and each class grows larger. When a class encompasses too many articles, the articles themselves are less similar, and we stop iterating. At this point, automatic classification is complete. Figure 14.4 tracks the clustering process from individual news articles to distinct categories. On the left, each point represents a news article. There are so many that they appear as a single dense block. After each iteration, the number of subclasses decreases, until we can pick them out clearly on the right.

In 1998, Florian and Yarowsky completed this work due to compelling necessity. That year, Yarowsky was the chairman of an international conference. He received hundreds of papers, all of which required peer review by specific experts in the field. To ensure that the papers would be justly reviewed, Yarowsky was to send them to the most authoritative figures in their respective domains. Although authors were required to define the scope of their papers, those fields

---

*Now a scientist at IBM Watson Laboratories.

[†]Radu Florian and David Yarowsky, *Dynamic nonlocal language modeling via hierarchical topic-based adaptation*, ACL 1999.

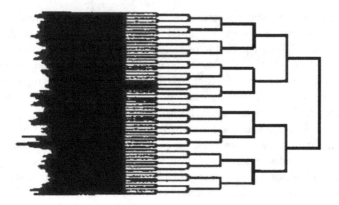

FIGURE 14.4: Text classification and aggregation, from Florian's own research.

were so broad they provided little guidance. A professor in his own right, Yarowsky did not have the time to sift through hundreds of papers himself. Instead, he devised a method of automatic classification, whose implementation was outsourced to his student, Florian. Thenceforth, that conference used the automatic classifier to distribute papers for the next few years. From this incident, we see a prevailing theme in Western engineering: we prefer to defer tasks to computers in place of manual labor. Though these methods require a short-term resources investment, they save time and money in the long run.

To wrap up, we have linked the law of cosines to news classification, an application we probably never foresaw when we learned the theorem in high school. As is the theme of this book, little mathematical tools have miraculously big impacts.

14.3   Extended reading: The art of computing cosines

Suggested background knowledge: numerical analysis

14.3.1   Cosines in big data

When we use Equation 14.2 to compute the angle between two vectors a, b, the asymptotic complexity is $O(|a| + |b|)$. If we assume that one vector is larger, $|a| > |b|$, then we can simplify the complexity to $O(|a|)$. Now suppose we have N articles. It takes $O(N \cdot |a|)$ to compute the angle between any vector and all others, and $O(N^2 \cdot |a|)$ to compute all pairwise distances in the dataset. Note that these bounds only apply to a single iteration. We can assume that the vocabulary size is 100,000 (so the feature vectors are very long), and that there are 10^{15} articles to be classified. If we own 100 servers, each of which can perform 100 million computations per second, then each iteration will take 100,000 seconds, or a little over a day. Dozens of iterations will take weeks, or even months— utterly impractical.

Having established this baseline, we can consider various speedups to this algorithm. Recall that Equation 14.2 computes two quantities: a dot product, and a product of magnitudes. The latter need not be computed from scratch each time; we can save the magnitudes of each vector and retrieve them when called for. For each pair of vectors, we are left with the dot product computation in the numerator. As a result, we can reduce the computation by approximately 2/3 per iteration, but we still have not changed the asymptotic complexity of this problem.

With regards to the dot product in Equation 14.2, we only consider non-zero elements of both vectors. If we represent the two vectors sparsely, then the computation complexity depends on the minimum number of non-zero elements between the two vectors. Generally, an article's length does not exceed 2,000, so the number of unique words hovers around 1,000. Thus, the computation complexity is reduced 100-fold, and each iteration's time drops an order of magnitude, from days to 15 minutes.

As a complement to vector sparsity, we can also remove stop words. In news classification, stop words include various functions of grammar, such as conjunctions ("and," "but"), prepositions ("in," "behind"), and adverbs ("very"). As we mentioned in the previous section, only similar news are alike; different news each contains its own features, which are unlikely to overlap. After removing stop words, we may still see many non-zero elements, but the common non-zero elements are vastly reduced. By skipping elements only non-zero in one vector, we can further reduce computation time by several times. As such, we can compute each iteration in a few minutes, and we will have completed the several dozen iterations by the end of day.

On a final note, stop-word removal not only improves the computation speed, but also increases the accuracy of classification. Since stop words are common among most articles, they only introduce noise to our classifier. We remove them for the same reasons that signal processing engineers filter out low frequency noise. From this, we again see the similarities between natural language processing and communications.

14.3.2 Positional weighting

As in search relevance, words that appear in different positions across a text do not carry the same importance. A headline obviously means more than a word hidden in the seventh paragraph of an article. Even in the main text, words at the beginning and end of each paragraph carry more meaning than words in the middle. In language arts classes throughout your childhood, your teacher must have emphasized this point. Special attention was given towards the introduction and conclusion, as well as the thesis and various topic sentences. The same ideas hold true in natural language processing. Therefore, we may introduce additional weights that consider the position of words in an article.

14.4 Summary

The methods introduced in this chapter produce high accuracy and performance, suitable for classifying texts on the order of millions. If we increase our data sizes to a hundred million, however, these computations are still quite slow. To tackle even larger datasets, we will introduce a faster, but less precise method in the next chapter.

Chapter 15

Solving classification problems in text processing with matrices

When I studied linear algebra in college, I did not know if it had any applications aside from solving linear equations. Many concepts that I encountered when learning about matrices, such as an eigenvalue, seemed to be particularly irrelevant to everyday life. Later, I enrolled in the numerical analysis course and learned approaches to solve matrix problems with computers, but I still could not think of any application for these methods beyond the classroom setting. I believe that most students who take these courses today still face the same confusion as I did, 30 years back. It was not until after many years of research in natural language processing that I started to realize that these ideas and algorithms from linear algebra had practical significance for real life.

15.1 Matrices of words and texts

Two of the most common classification problems in natural language processing are classifying articles by topic (e.g., classifying all news about the Olympics as athletics) and classifying words by meaning (e.g., classifying the terms of all the sports as athletics). Both of these problems can be gracefully solved with matrix computations. In order to illustrate this, we shall take a look at our examples from the previous chapter, the law of cosines and news classification.

News classification, or any text classification, is a cluster problem that requires computing the degree of similarity between two pieces of news. This process starts with converting the news into content words that convey meanings. We then transform the words into a series of numbers, or a vector to be precise, and finally calculate the angle between the two vectors. A small angle means the two pieces of news are highly similar, while a large angle between two feature vectors signifies a big difference in topics of discussion. Of course, when the two vectors are orthogonal to each other, the two articles are unrelated. This algorithm is clean in theory, but requires multiple iterations because it makes comparisons in pairs. Especially when dealing with many pieces of news and a huge lexicon, this approach would be extremely time-consuming. Ideally, our solution would calculate the correlation between all pieces of news at once. This solution uses Singular Value Decomposition, or SVD, a matrix factorization method.

Now, let us turn our attention to SVD. First, we need to use a large matrix to describe thousands of pieces of news and millions of word correlations with each other. Every row of this matrix corresponds to a news article, and every column to a word. If we have N words and M pieces of news articles, we end up with the following $M \times N$ matrix:

$$A = \begin{bmatrix} a_{11}...a_{1j}...a_{1N} \\ .... \\ a_{i1}...a_{ij}...a_{iN} \\ .... \\ a_{M1}...a_{Mj}...a_{MN} \end{bmatrix} \tag{15.1}$$

The element of row i and column j, a_{ij}, is the weighted word frequency of the jth word in the lexicon's appearance in the ith news article. Needless to say, this matrix is enormous. For example, if $M = 1,000,000$ and $N = 500,000$, there would be 500 billion elements - about 2.5 square miles if printed out in size 10 font!

SVD decomposes a large matrix, like the one mentioned above, into the product of three smaller matrices, as shown in the figure below. Matrix X is a million by a hundred, matrix B is a hundred by a hundred, and matrix Y is a hundred by half a million. The total number of elements in the three matrices is no more than 150 million, which is less than 1/3000 of the original matrix in size. The corresponding storage capacity and computation capacity would also decrease by more than three orders of magnitude. See Figure 15.1.

The three matrices have corresponding physical meanings. The first matrix X is the result of word classification. Every row of matrix X represents a word, every column a semantic class where words with similar meanings are grouped together. Every non-zero element in this row represents the word's importance (or correlation) in each semantic class. A larger value means greater correlation.

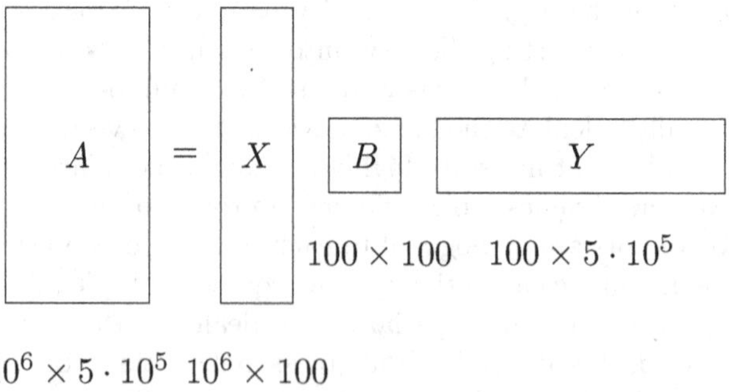

$$A = X \quad B \quad Y$$

$$100 \times 100 \quad 100 \times 5 \cdot 10^5$$

$$10^6 \times 5 \cdot 10^5 \quad 10^6 \times 100$$

FIGURE 15.1: A large matrix can be decomposed into three smaller matrices.

Taking this small 4×2 matrix as an example:

$$X = \begin{bmatrix} 0.7 & 0.15 \\ 0.22 & 0.49 \\ 0 & 0.92 \\ 0.3 & 0.03 \end{bmatrix} \tag{15.2}$$

There are four words and two semantic classes in this matrix. The first word is highly related to the first semantic class (the correlation is 0.7), and not so related to the second semantic class (the correlation is 0.15). The second word has the opposite pattern. The third word is only related to the second semantic class, and not at all related to the first one. The last word is not exactly related to either semantic class, because both elements in that row (0.3 and 0.03) are small. Relatively speaking, it is more related to the first one than to the second.

The last matrix Y is the result of text classification. Each column in matrix Y corresponds to an article, and each row to a topic. Each element in the column represents the article's correlation to a different topic. Here's an example in a 2×4 matrix:

$$Y = \begin{bmatrix} 0.7 & 0.15 & 0.22 & 0.39 \\ 0 & 0.92 & 0.08 & 0.53 \end{bmatrix} \tag{15.3}$$

This matrix has four articles and two topics. The first article obviously fits under the first topic. The second article is highly related to the second topic (the correlation is 0.92), but it is also slightly related to the first topic (0.15). The third article is not too related to either, but relatively speaking it is closer to the first topic. The fourth article is somewhat related to both, with a relatively higher correlation to the second topic. If we only keep the largest value in each column, then each article would be classified into only one topic: the first and third articles belong to the first topic, the second and fourth articles to the second topic. This outcome is similar to what we obtained through using the law of cosines for classification, in which every article is classified into one topic.

The middle matrix B represents the correlation between the classes of words and those of articles. Take the following 2×2 matrix as an example:

$$B = \begin{bmatrix} 0.7 & 0.21 \\ 0.18 & 0.63 \end{bmatrix} \tag{15.4}$$

In this matrix B, the first semantic class is related to the first topic and not so much to the second, and the opposite is true for the second semantic class.

Hence, through SVD, both term classification and text classification can be accomplished at once. Moreover, the result even tells us the correlation between each semantic class and each topic. What a beautiful solution!

The only problem left for now is to teach computers to implement SVD. Many concepts in linear algebra, such as eigenvalues and numerical analysis algorithms, can play a big role in this step. While smaller matrices can be computed with programs like MATLAB, larger matrices, which can be as large as millions by millions in dimension, require multiple computers working in conjunction to process. Because it is difficult to break down SVD into irrelevant sub-operations, even Google could not decompose matrices with computational advantage in the past, despite having parallel computing tools like MapReduce. In 2007, Dr. Edward Chang from Google China led several Chinese engineers and trainees to implement the parallel algorithm for SVD. This is one of Google China's contributions to the world.

15.2   Extended reading: Singular value decomposition method and applications

Suggested background knowledge: linear algebra

In this chapter, we introduced the method of singular value decomposition. Mathematically, it is defined as the decomposition of matrix A into the product of three matrices:

$$A_{MN} = X_{MM} \times B_{MN} \times Y_{NN} \tag{15.5}$$

Here, X is a unitary matrix and Y is the conjugate transpose of a unitary matrix. A matrix is unitary if its conjugate transpose is also its inverse, and their product is the identity matrix. Hence, both the unitary matrix and its conjugate transpose are square matrices. Matrix B is a diagonal matrix in which the entries outside the main diagonal are all zero. Here's an example from Wikipedia. Let

$$A = \begin{bmatrix} 1 & 0 & 0 & 0 & 2 \\ 0 & 0 & 3 & 0 & 0 \\ 0 & 0 & 0 & 0 & 0 \\ 0 & 4 & 0 & 0 & 0 \end{bmatrix} \tag{15.6}$$

We can decompose A as

$$A = \underbrace{\begin{bmatrix} 0 & 0 & 1 & 0 \\ 0 & 1 & 0 & 0 \\ 0 & 0 & 0 & -1 \\ 1 & 0 & 0 & 0 \end{bmatrix}}_{B} \underbrace{\begin{bmatrix} 4 & 0 & 0 & 0 & 0 \\ 0 & 3 & 0 & 0 & 0 \\ 0 & 0 & \sqrt{5} & 0 & 0 \\ 0 & 0 & 0 & 0 & 0 \end{bmatrix}}_{X} \underbrace{\begin{bmatrix} 0 & 1 & 0 & 0 & 0 \\ 0 & 0 & 1 & 0 & 0 \\ \sqrt{0.2} & 0 & 0 & 0 & \sqrt{0.8} \\ 0 & 0 & 0 & 1 & 0 \\ -\sqrt{0.8} & 0 & 0 & 0 & \sqrt{0.2} \end{bmatrix}}_{Y}.$$

$$\tag{15.7}$$

It is clear that both X and Y are unitary matrices.

You might notice that SVD did not reduce the order of magnitude for our matrices above, as it did in the last section's example. However, since many values in the diagonal matrix B are minuscule or even zero, they can be omitted. Thus, a large matrix becomes the product of three smaller matrices through SVD, consistent with our conclusion in the last section.

SVD is typically carried out in two steps. First, we need to transform matrix A into a double diagonal matrix where all values but those on the main diagonal are zero. The computation for this step is $O(MN^2)$, assuming $M > N$. We could also use matrix A's sparsity to reduce computation time. The second step is decomposing the double diagonal matrix into three matrices. The amount of computation required for this step is relatively small compared to that for the first step.

In text classification, M represents the number of articles while N corresponds to the size of the lexicon. While the time required for article similarity computation using SVD and that for one iteration of computation using the law of cosines are on the same order of magnitude, the former is much less time-consuming because it does not involve multiple iterations. One problem with SVD is the amount of storage it requires for the entire matrix, which is not necessary for clustering using the law of cosines.

15.3   Summary

Compared with the use of text features, vectors, and cosine that we saw in the previous chapter, singular value decomposition has the practical advantage of faster computation because of its single iteration. However, the result obtained from this method is not as refined, which makes it more suitable for computing rough classifications of articles on a large scale. In practical application, we can obtain a rough result first by using SVD, then compute more precise results through multiple iterations of the dot product method. A combination of these two methods gives us the best of both worlds, allowing us to be more efficient without losing precision.

Bibliography

1. Bellagarda, J.R. "Exploiting latent semantic information in statistical language modeling," Proceedings of the IEEE, Volume:88 Issue:8, 1279–1296, August 2008.

Chapter 16

Information fingerprinting and its application

16.1 Information fingerprint

Previously, we discussed that the amount of information in the content of text is its information entropy. Therefore, the shortest length of the lossless encoding of an information source cannot be smaller than its entropy. Granted, the actual compressed length is always somewhat longer than its information entropy in number of bits. However, if we only need to distinguish two sections of text or two images, we can use a code shorter than its entropy. Any piece of information, including text, audio, video, or image, can be assigned a random number that serves as its information fingerprint. This information fingerprint allows us to distinguish one piece of information from another. As long as the fingerprinting algorithm is well designed, two distinct pieces of information will not share the same fingerprint, just like how human fingerprints are rarely identical. Information fingerprint is widely applicable in information encryption, compression, and processing.

In Chapter 9 on graph theory and web crawlers, we mentioned that the URL needs to be stored in a hash table in order to avoid repeated visits to the same webpage. However, storing URLs in strings takes up unnecessary space and wastes searching time, which becomes even more cumbersome as URLs tend to be very long nowadays. For example, if you were to Google "Beauty of Mathematics Jun Wu," the corresponding URL is longer than a hundred characters:

> https://www.google.com/search?q=beauty+of+mathematics+jun
> +wu\&oq=beauty+of+mathematics+jun+wu\&aqs=chrome..69i
> 57.5760j0j7\&sourceid=chrome\&ie=UTF-8

In 2010, the total number of webpages has reached the order of magnitude of 500 billion. Suppose that the average length of a URL is 100 characters, then storing 500 billion URLs would require 50TB, which amounts to 50 million megabytes. If we consider the fact that hash table typically has the storage efficiency of 50%, then the actual storage needed would be over 100TB. The typical server in 2010 has the storage capacity of 50GB, requiring about 2,000 servers to store all the URLs. Even if we manage to store everything, searching in the format of strings is highly inefficient due to the unequal lengths of URLs.

Therefore, if we could find a function that randomly maps the 500 billion URLs onto 128-bit binaries, which is 16 bytes, the above string would correspond to the following random number:

8932494329843984329805454545543

Thus, each URL would occupy 16 bytes of space rather than the original 100 bytes. This method reduces the memory requirement for storing URLs to less than 1/6 of the original. This 16-byte random number is the information fingerprint for this particular URL. It can be proven that as long as the random number generating algorithm is sufficient, it is almost impossible to have duplicate fingerprints for different strings of characters, just as two different people cannot have the same fingerprints. Because all information fingerprints are standardized to be 128-bit binary integers, the amount of computation involved in searching is greatly reduced compared to that required for strings. When web crawlers download webpages, they index all visited URLs into hash tables in the form of information fingerprints. Whenever they encounter a new webpage, the information fingerprint for that URL is computed and compared to existing fingerprints in the hash table to determine whether to download this webpage. This kind of integer search is multitudes faster than a string search. For some web crawlers, 64-bit binary fingerprints would suffice, which can save storage space and computational time even further.

The process of calculating the information fingerprint of the Google URL (string) above is generally divided into two steps. First, we need to treat the string as a special integer that is extraordinarily long. This is relatively easy because all strings are stored as integers in the computer. The next step requires a key algorithm for producing information fingerprints: a Pseudo-Random Number Generator (PRNG). Through PRNG, any long integers can be transformed into a pseudo-random number of a set length. The PRNG algorithm was first proposed by John von Neumann, the founding figure in computing. His method was fairly simple: delete the beginning and the end of a number and keep just the middle digits. For instance, the square of the four-digit binary number 1001 (9 in decimal) is 0101001 (81 in decimal), and it would become 0100 after having its beginning (01) and end (1) taken out. Of course, the numbers produced through this method are not entirely random, which means that two different pieces of information could potentially have the same fingerprint. The number 0100 (4 in decimal) would also have the fingerprint of 0100 when processed this way. Nowadays, the commonly used Mersenne Twister algorithm generates much better random numbers.

Information fingerprinting has many applications beyond detecting and eliminating duplicate URLs. Its twin brother is the password. One of the key features of information fingerprinting is its irreversibility. In other words, it is impossible to access an information source from its fingerprint. This characteristic is precisely what is needed in encrypted network transmission. For example, a cookie is a type of information fingerprint, and a website can use a user's local client cookie to identify different users. Websites cannot access the user's personal information through this fingerprint, which protects the user's privacy. However, the cookie itself is usually not encrypted, so it is possible to figure out which computer accessed which websites through analyzing the cookies. In order to ensure information security, some websites (like bank websites) use encrypted HTTPS and encrypt the users' cookies when they visit the websites. The reliability of encryption depends on how difficult it is to access information with the same fingerprint, such as whether a hacker can reproduce a certain user's cookie. From the perspective of encryption, the algorithm of Mersenne Twister can still be improved. Since the random numbers generated still have a degree of correlation, cracking one means cracking many.

Encrypting on the internet uses a encryption-based Cryptographically Secure Pseudo-Random Number Generator, or CSPRNG. Common algorithms include MD5 and SHA-1, which can transform information of different-lengths into set-length 128-bit or 160-bit binary random numbers. It is worth mentioning that the SHA-1 algorithm used to be considered flawless, but Professor Xiaoyun Wang from China proved that it actually has loopholes. However, there is no need to panic; this does not necessarily mean that hackers can actually access your information.

16.2   Applications of information fingerprint

Although information fingerprinting has had a long history, it did not become widely used until after the advent of the internet in recent decades.

In the last section, we talked about how web crawlers can use information fingerprints to determine whether a webpage has already been downloaded quickly and economically. Information fingerprinting has a lot of applications in natural language processing and the internet that are too many to enumerate. We only need to look at a few representative examples here.

16.2.1   Determining identical sets

During a webpage search, sometimes one needs to determine whether the terms used in two searches are identical (although the order of terms might differ), such as "California, Mountain View, Starbucks" and "Starbucks, California, Mountain View." A more common way to express this problem is determining whether two sets are identical (for example, whether a person is using two different email accounts to send junk emails to the same group of recipients). There are many ways to solve this problem. None of the solutions are absolutely right or wrong, but some are better and more efficient than others.

The most cumbersome method is to compare all of the elements in the set one by one. The time complexity of calculation by this method is $O(N^2)$, where N is the size of the set. If someone were to give me this answer in an interview, that person would not pass.

A slightly better approach is to sort the elements in the two sets respectively and make ordered comparisons. The time complexity for this method is $O(N \log N)$, which is a big step up from the first approach but still not good enough. A similar solution is to put the first set in a hash table and compare every element in the second set to the hash table elements. This method has the time complexity of $O(N)^*$ but it requires extra space of $O(N)$ and has overly complex code. None of these methods are ideal.

The best solution is to calculate the information fingerprints of the two sets and then make a direct comparison. For a set $S = \{e_1, e_2, \cdots, e_n\}$, we define its fingerprint as $FP(S) = FP(e_1) + FP(e_2) + \cdots + FP$. Here, $FP(e_1), FP(e_2), \ldots, FP(e_n)$ are respective fingerprints of the elements in S. Because of the commutative property of addition, we can be sure that a set's fingerprint would not change if its elements were to appear in a different order. If two sets have the same elements, then they must have the same fingerprint. Of course, the probability of different elements having the same fingerprint is minuscule and practically negligible. We will discuss how small this probability is in the extended reading section.

Because the complexity for computing using information fingerprints is $O(N)$ and no extra space is required, this is the ideal method.

Information fingerprinting has many other similar applications. For example, in order to determine whether a song from the internet is a pirated duplicate, we only need to compute the information fingerprints for the two audio files and compare them.

16.2.2   Detecting similar sets

The thoughtful reader might raise the following challenge: why would the sender of junk mail be so stupid as to include the exact same recipients when sending from two different accounts? If one or two recipients were different, then the above method would no longer work. In order to solve this problem, we have to be able to quickly determine whether two sets are basically the same, or highly similar. To do this, we only need to make slight modifications to the method above.

We could randomly select several recipients from the email address lists of two accounts following the same rules, say the addresses that end with the number 24. If they have the same fingerprints, then it is very likely that the two email address lists are almost identical. Because only a limited number of addresses are selected, usually in the single digit, it is easy to determine whether the two sets are 80% or 90% identical. '

---

*The time complexity of $O(N)$ cannot be overcome because all of the N elements need to be looped through.

The algorithm for determining similar sets introduced above has many practical applications. For example, when browsing webpages, it can be used to determine whether two pages are identical. The computation time for calculating all of the content of the two webpages would be unnecessarily long. We only need to select several terms from each webpage that constitute a set of feature terms, then compute and compare the information fingerprints of the feature sets. Since common words would generally appear in two webpages being compared, they are unsuitable for the feature set. The words that only appear once, a.k.a. singletons, could be noise and should not be considered as feature terms either. Among the remaining terms, we only need to locate several with the largest inverse document frequency and compute their information fingerprints because we know that terms with large inverse document frequency are highly identifiable. If two webpages have the same information fingerprint computed according to this approach, then they are almost identical. In order to accommodate for a certain margin of error, Google employs a special information fingerprint algorithm called SimHash (a compound word of "similar" and "hash"). We will explain the principle of SimHash in the extended reading section.

We can also use the above method with minor modifications to determine whether an article is a plagiarized version of another one. Specifically, we need to cut each article into smaller pieces and select a set of feature terms for each piece, then compute their information fingerprints. By comparing these fingerprints, we can locate large chunks of identical texts and determine which one is the original and which one is plagiarized based on their times of creation. Google Labs created a project called CopyCat using this method, which can accurately locate the original article and its copier.

16.2.3   YouTube's anti-piracy

Google's YouTube is the world's largest video-sharing website, where all contents are uploaded by its users. These users include professional media organizations such as NBC and Disney, as well as personal users. Because of the relative lack of restrictions for the latter group, some users upload content that belongs to professional media organizations. This problem, if left unsolved, would be detrimental to the survival of YouTube.

It is not a simple task to determine whether a certain video is a pirated version of another video among millions of videos. The size of a compressed video of several minutes could range from a few megabytes to a dozen megabytes. If restored to images of 30 frames per second, the amount of data would be enormous. Therefore, no one determines whether two videos are identical by direct comparison.

The two key techniques of video matching are main frame extraction and feature extraction. MPEG videos (displayed on NTSC monitors) have 30 frames per second, but there is little difference between one frame and the next (otherwise the video would be incoherent). Only a few frames are complete images, which we call main frames. The rest of the frames merely store difference values compared to these main frames. The importance of main frames to videos is comparable to the centrality of keywords in news. Hence, video image processing starts with locating the main frames, followed by representing these main frames with a set of information fingerprints.

With the information fingerprints calculated, the anti-piracy problem becomes analogous to determining whether the elements in two sets are identical. After Google bought YouTube, image processing scientists at the Google Research developed an anti-piracy system that works very well. Because of its ability to locate identical original videos and their copies, Google came up with a very interesting advertising revenue share strategy: although all videos can embed advertisements, only the original videos are entitled to advertisement revenue even if the advertisement is embedded in the pirated video. This way, none of the websites that copy and upload other people's videos can gain profit from doing so. Without economic profit, the occurrence of pirating decreased significantly.

16.3   Extended reading: Information fingerprint's repeatability and SimHash

Suggested background knowledge: probability theory and combinatorics

16.3.1   Probability of repeated information fingerprint

Information fingerprints are generated as pseudo-random numbers. Since they are pseudo-random, two different pieces of information could potentially produce the same fingerprint. Although this probability exists theoretically, it is extremely small. We will discuss exactly how small it is in this section.

Suppose the production of pseudo-random numbers ranges from 0 to $N-1$, and there are N numbers in total. For 128-bit binaries, $N = 2^{128}$, which is an enormous number. If we were to choose two fingerprints randomly, the probability of them being repeated is $1/N$, and the probability of no repeat is $(N-1)(N)$, because the first number could be any number and the second one is chosen out of the remaining $N-1$ numbers. If we were to choose three numbers without repeating, then the third number has to be selected from one of $N-2$, which makes the probability of no repeat $(N-1)(N-2)/N^2$. Following the same logic, the probability of non-repeated k fingerprints is

$$\frac{(N-1)(N-2)\dots(N-k+1)}{N^{k-1}}.$$

P_k becomes smaller as k becomes larger: when the number of fingerprints produced is big enough, there is a possibility of repetition. If $P_k < 0.5$, then the mathematical expectation of having one repeat in k fingerprints is greater than 1. Now, let us estimate the maximum value of k at this moment.

The above probability is equivalent to the following:

$$P_{k+1} = \frac{(N-1)(N-2)\ldots(N-k)}{N^k} < 0.5 \tag{16.1}$$

when $N \to \infty$:

$$P_{k+1} \approx e^{-\frac{1}{n}} e^{-\frac{2}{n}} \ldots e^{-\frac{k}{n}} = \exp\left[-\frac{k(k+1)}{2N}\right] \tag{16.2}$$

This probability needs to be smaller than 0.5, therefore:

$$P_{k+1} \approx \exp\left[-\frac{k(k+1)}{2N}\right] < 0.5 \tag{16.3}$$

which is equivalent to:

$$k^2 + k_2 N \log 0.5 > 0.5 \tag{16.4}$$

Because $k > 0$, the above inequality has a unique solution:

$$k > \frac{-1 + \sqrt{1 + 8N \log 2}}{2} \tag{16.5}$$

This means that for a very large N, k is a very large number. If we were to use MD5 fingerprint (though it has its flaws), it has 128-bit binary and $k > 2^{64} \approx 1.8 \times 10^{19}$. In other words, there would only be a repeated fingerprint out of every 180 trillions. Hence, the probability of different information having the same fingerprint is almost zero. Even if we use 64-bit fingerprints, the probability of repeat is still minuscule.

16.3.2 SimHash

SimHash is a special kind of information fingerprint, created by Moses Charikar in 2002. However, the first time it gained wide recognition was not until Google Crawler used it for webpage duplicate detection and published the results in the Proceedings of the 16th International Conference on World Wide Web. Although Charikar's paper is rather obscure, the principle of SimHash is not too complicated to understand. We shall illustrate this with an example of Google Crawler's duplicate detection during webpage downloads.

Suppose there are a certain number of terms in a webpage, $t_1, t_2, \ldots, t_k$, and their weights (for example, their term frequency/inverse document frequency [TF-IDF] value) are $w_1, w_2, \ldots, w_k$. For the purpose of this illustration, let us suppose that these terms have 8-bit binary information fingerprints, although they would be much longer in real life because of high repeatability. The calculation of SimHash is divided into two steps.

I call the first step expansion, which involves expanding the 8-bit binary fingerprints into 8 real numbers, represented by $r_1, r_2, \ldots, r_8$. The values of these real numbers are calculated as follows:

First, we set their initial values to 0. Then, looking at the 8-bit fingerprint of t_1, if the ith digit of t_1 is 1, then we add w_1 to r_i; if it is 0, then we subtract w_1 from r_i. For example, suppose the fingerprint of t_1 is 10100110 (a random number), the values of r_1 through r_8 would be the following after being processed with t_1:

TABLE 16.1: The values of r_1 through r_8 after processing the first term.

$r_1 = 1$	w_1
$r_2 = 1$	$-w_1$
$r_3 = 1$	w_1
$r_4 = 1$	$-w_1$
$r_5 = 1$	$-w_1$
$r_6 = 1$	w_1
$r_7 = 1$	w_1
$r_8 = 1$	$-w_1$

Now, let us look at the second term with the fingerprint 00011001. Following the same principle of adding if it is 1, and subtracting if 0, we know that the weight of t_2, which is w_2, should be subtracted from r_1 because the fingerprint's first digit is 0. This way, $r_1 = w_1 - w_2$, and we obtain the following table of values for $r_2, \ldots, r_8$:

TABLE 16.2: The values of r_1 through r_8 after processing the first two terms.

r_1	$w_1 - w_2$
r_2	$-w_1 - w_2$
r_3	$w_1 - w_2$
r_4	$-w_1 + w_2$
r_5	$-w_1 + w_2$
r_6	$w_1 - w_2$
r_7	$w_1 - w_2$
r_8	$-w_1 + w_2$

After going through all of the terms, we end up with 8 final numbers $r_1, \ldots, r_8$, which marks the end of expansion. Suppose the values of $r_1, r_2, \ldots, r_8$ are as follows after expansion:

TABLE 16.3: The values of r_1 through r_8 after processing all of the terms, turning positive numbers to 1 and negative numbers to 0.

r_1	−0.52	0
r_2	−1.2	0
r_3	0.33	1
r_4	0.21	1
r_5	−0.91	0
r_6	−1.1	0
r_7	−0.85	0
r_8	0.52	1

I call the second step contraction, which is transforming 8 real numbers back into an 8-bit binary. This is a very simple process. If $r_i > 0$, then set the corresponding digit of the binary to 1. Otherwise, set it to 0. Now, we have obtained the 8-bit SimHash fingerprint of this article. For the above example, the article's SimHash = 00110001.

A feature of SimHash is that a smaller difference between the SimHash of two webpages indicate a higher degree of similarity. If two pages are duplicates, then they must have identical SimHash. If they are almost identical and differ only in a few words with small weights, then we can be sure that they will also have the same SimHash. It is worth mentioning that if two webpages have different but highly similar SimHash, they would also be highly similar. When making comparisons with 64-bit SimHash, the difference in one or two digits means that the probability of duplicate contents in the two corresponding webpages is greater than 80%. By recording the SimHash of every webpage and then checking whether the SimHash of a new page has already appeared, we can detect duplicates and avoid wasting computer resources on redundant index construction.

Because of its simple principle and ease of application, information fingerprinting is used in many fields. It is an especially indispensable tool when working with today's massive amounts of data.

16.4 Summary

Information fingerprints can be understood as a random mapping of a piece of information (text, image, audio or video file) onto a point in multidimensional binary space (a binary number). As long as the random number generator function is well designed, different points corresponding to difference pieces of information will not overlap and these binary numbers can constitute unique fingerprints of the original information.

Bibliography

1. Charikar, Moses. "Similarity estimation techniques from rounding algorithms," Proceedings of the 34th Annual ACM Symposium on Theory of Computing. 2002.

2. Gurmeet Singh, Manku; Jain, Arvind; Das Sarma, Anish. "Detecting near-duplicates for web crawling," Proceedings of the 16th International Conference on World Wide Web. 2007.

Chapter 17

Thoughts inspired by the Chinese TV series Plot: The mathematical principles of cryptography

In 2007, I watched the Chinese TV series *Plot (An suan)* and was intrigued by its storyline and the acting. One of the stories in the series relates to cryptography. I thought the story itself was not bad, notwithstanding a certain degree of deliberate mystification. The show did get one thing right: mathematics is the foundation of modern cryptography.

17.1 The spontaneous era of cryptography

The history of cryptography can be traced back to more than two thousand years ago. It was recorded that Julius Caesar sent messages using secret codes in order to prevent them from being intercepted by the enemy. Caesar's method was very simple: each plaintext letter in the alphabet was mapped to a corresponding ciphertext letter, as shown in Table 17.1:

TABLE 17.1: Corresponding plaintext and ciphertext table of Caesar's cipher.

plaintext	ciphertext
A	B
B	E
C	A
D	F
E	K
...	...
R	P
S	T
...	...

Now, without the cipher table, one would not be able to understand anything even if the information was intercepted. For example, CAESAR would be coded as ABKTBP, which is a nonsense word to the enemy. This encryption technique is known as "Caesar's cipher." There are Caesar's cipher toys on the market, like the one shown in Figure 17.1:

FIGURE 17.1: A Caesar's cipher toy.

Of course, anyone with a background in information theory would know that as long as more intelligence is gathered (even if the information is encrypted), the code can be cracked by computing the frequency of letters. Sir Arthur Conan Doyle introduces this technique in "The Adventure of the Dancing Men," a story from *The Return of Sherlock Holmes* (see Figure 17.2). In recent years, many screenwriters of spy-themed Chinese TV shows still use this kind of bad cipher. For example, the price of groceries (a series of numbers) would correspond to the page and index of the word in a dictionary. For people who have studied information theory, deciphering this kind of code does not even require a code-book - collecting intelligence a few more times would suffice.

From the time of Caesar to the beginning of the 20th century, cryptographers improved their techniques at a slow pace because their work relied mostly on experience and intuition, rather than on mathematical principles (of course, information theory had not been in existence back then). People gradually realized that a good encoding method should make it impossible for the decryptor to obtain the statistical information of the plaintext from the ciphertext. Experienced encryptors encoded common words into multiple corresponding ciphertexts, thus making it harder for the decryptor to follow any pattern. For example, if the word "is" were coded only as 0543, then the decryptor will notice that 0543's frequency of appearance is very high. However, if the word were to correspond to 0543, 373, 2947, and seven other codes, and each time a random code was chosen to represent "is," then

FIGURE 17.2: The dancing men: they seem mysterious but are in fact easy to decipher.

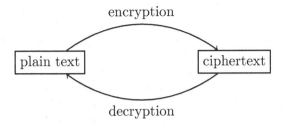

FIGURE 17.3: Encryption and decryption are a pair of function and inverse function.

none of the codes would have a frequency that is too high. Moreover, there is no way for the decryptor to know that all of them correspond to the same word. This technique already follows some simple principles of probability theory.

Good encryption makes it impossible to infer new ciphertexts from a known pair of ciphertext and plaintext. From a mathematical standpoint, the encryption process can be seen as the operation F of a function, and the decryption process is the operation of the inverse function, as shown in Figure 17.3. Plaintext is the independent variable, and ciphertext is the function value. A good encryption function cannot be deduced through several independent variables and function values. However, encryptors did not abide by this rule with too much success before WWII. Historically, there had been many instances of ill-designed encryption systems. For example, the Japanese military's cryptography was especially problematic during WWII. The US was able to decipher many Japanese ciphertexts. Before the Battle of Midway, the US had started to notice that the intercepted Japanese cipher telegrams often mentioned a place coded as AF, an island on the Pacific, but the identity of this island was unknown. Hence, the US started to release fake news about each of the islands under its control respectively. When the fake news of "Midway water purification system breakdown" was released, the US intercepted new Japanese cipher telegram containing AF (the telegram's content was "AF is short on water"). Thus, the US determined that Midway was the island that AF referred to, and this was confirmed by what happened later when the US successfully ambushed the Japanese Combined Fleet.

The late American cryptanalyst Herbert Osborne Yardley (1889-1958) had helped the Chinese government decode the Japanese cipher in Chongqing during WWII. His greatest success during those two years was cracking the communication code used between the Japanese forces and the spies in Chongqing. This helped decrypt thousands of telegrams exchanged between the Japanese army and the spies, which contributed to solving the spy case of Kuomintang traitor "One-Armed Bandit" who was supplying Chongqing weather information to the Japanese army. The work accomplished by Yardley (along with a Chinese woman named Xu Zhen) significantly mitigated the damage caused by the Japanese bombing of Chongqing. After returning to the US, Yardley wrote a book about this experience called *The Chinese Black Chamber*, which was not authorized to declassify and publish until 1983. This book shows us that the Japanese army's cryptography had serious drawbacks. The codebook used by the Japanese army and the Chongqing spies was *The Good Earth* by famous American author Pearl S. Buck, winner of the Nobel Prize for Literature in 1938. This book was easy to access, and the decryptor only needed to have this book at hand to decode cipher telegrams. The page number of the cipher is a very simple formula: the month plus the day of the month of the telegraph date plus 10. For example, if the telegram was sent on March 11, then the page number is $3 + 11 + 10 = 24$. This kind of cryptography breaks our aforementioned rule of "deducing encryption function through several independent variables and function values." For a system like this, deciphering one cipher telegram makes it possible to decipher all subsequent ones.

Yardley's book also mentions that the Japanese army had limited knowledge of the technical principles of encryption. Once when the Japanese embassy at Manila was sending a telegram, the machine froze halfway through so the exact same telegram was sent again. This kind of same-text cipher telegram is a cryptographical taboo (similar to the security key we use to log in with virtual private network [VPN], the cipher machine's rotor disks should automatically rotate during encryption to ensure that the same key is not used repeatedly, which means that even the same telegram should be encrypted differently for the second time). Moreover, when the Ministry of Foreign Affairs of Japan was updating its cipher machines, some embassies in countries farther away were still using the old machines because the arrival of the new ones had been delayed. This created a situation where two generations of cipher machines were in use simultaneously, enabling the US to intercept two sets of cipher telegrams from both the old and the new machines. Since a large proportion of the old Japanese code had already been deciphered, this undermined the secrecy of the new cipher. To sum up, Japanese intelligence during WWII was often intercepted and deciphered by the Americans. The famous Japanese Navy Marshal Admiral Isoroku Yamamoto was killed because of it.* We know that falling behind leaves one vulnerable to attacks; in fact, not knowing how to use math leaves one vulnerable to attacks too.

17.2 Cryptography in the information age

During WWII, many top scientists including Claude E. Shannon who made theoretical contributions to information theory had worked for the US

---

*The US obtained information regarding the Yamamoto plane's whereabouts through deciphering Japanese intelligence and shot it down with a fighter.

intelligence agency. Information theory is the direct product of information science. After Shannon's proposal of information theory, cryptography entered a new era. According to information theory, the ideal cipher does not provide the enemy any additional information even after it is intercepted. In information theory jargon, the amount of information available to the attacker should not increase. Generally, when ciphertexts are evenly distributed and statistically independent, the minimal amount of information is provided. An even distribution prevents the attacker from performing statistical analyses, while statistical independence prevents chosen-plaintext attacks even if the attacker obtains a plaintext and its corresponding ciphertext. In the TV series, *Plot*, codebreaker Chen who successfully deciphers one cipher telegram following the traditional method is unable to generalize it. On the other hand, the mathematician codebreaker Huang in *Plot* foresees this problem because she knows that the enemy's new cryptography system produces statistically independent ciphertexts.

With information theory, cryptography is grounded in a theoretical foundation. The now commonly used public key cryptography, the "Back to the Mainland" system mentioned in *Plot*, is based on this theory.

The principle of public key cryptography is in fact quite simple. We shall encrypt and decrypt the word Caesar to illustrate this. First, we need to turn the word into a series of numbers, and use this as our plaintext. For example, Caesar's ASCII Code $X = 067097101115097114$ (each letter is represented by three digits), and use this as our plaintext. Now, we are ready to design a cryptography system to encrypt this plaintext.

1. Find two large prime numbers, P and Q, the bigger the better, such as 100-digit long ones. Then, calculate their product.

$$N = P \times Q$$
$$M = (P - 1) \times (Q - 1)$$

2. Find integer E, co-prime with M. That is, M and E have no common denominator other than 1.

3. Find integer D so that the remainder of $E \times D$ divided by M is 1, i.e., $E \times D \mod M = 1$.

Now, we have designed the most advanced and most commonly used cryptography system. Here, E is the public key that anyone can use to encrypt and D is the private key for decryption, which we should keep to ourselves. The product N is public, and it is alright if this is known by the enemy.

We can now use the following equation to encrypt X to obtain ciphertext Y.

$$X^E \mod N = Y \tag{17.1}$$

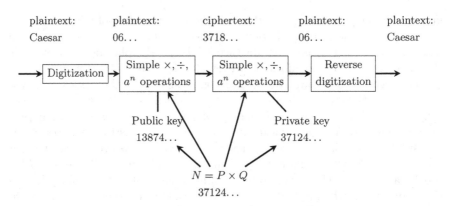

FIGURE 17.4: Public key diagram.

Now, without security key D, no one can restore X from Y. According to Fermat's little theorem,* if D is known, then we can easily obtain X from Y by the following equation:

$$Y^D \mod N = X \qquad (17.2)$$

This process can be summarized by Figure 17.4.

The advantages of public key encryption include:

1. Simplicity. It is just some multiplication and division arithmetic operations.

2. Reliablity. The public key method ensures that ciphertexts thus produced are statistically independent and evenly distributed. In other words, no matter how many plaintexts and corresponding ciphertexts are given, one cannot decipher a new ciphertext with the known ones. Most importantly, N and E are public and can be used by anyone to encrypt, but only the person with the key D at hand is able to decrypt, which is to say that even the encryptors cannot decrypt. This way, even if the encryptor is caught by the enemy and becomes a traitor, the whole encryption system is still safe. (In contrast, for Caesar's cipher, if someone who knows the codebook leaks it, the entire cipher system is made public.)

3. Flexibility. It can generate many combinations of public key E and private key D for different encryptors.

---

*Fermat's little theorem has two equivalent descriptions. Description 1: P is a prime number; for any integer N, if P and N are coprime, then $N^{P-1} \equiv 1 \pmod{P}$. Description 2: P is a prime number; for any integer $N^P \equiv N \pmod{P}$.

Lastly, let us look at the difficulty to decipher this kind of code. First, we need to establish that there is no cipher in the world that cannot ever be deciphered, but the difference is how long it stays effective. Research up to this date suggests that the best way to solve the public key encryption algorithm is by factoring the large number N, which is to find P and Q through N. However, the only way to find P and Q at present is by trying every number one-by-one using the computer. The success of this cumbersome approach depends on the computation power of the computer, which is why both P and Q need to be very large. An encryption method can be considered satisfactory if it ensures security against decryption in 50 years. The RSA-158 cipher that was decrypted some years ago was thus factored:

$$39505874583265144526419767800614481996020776460304936454139376051579355626529450683609727842468219535093544305870490251995655335710209799226484977949442955603 = 33884958374667213934368393204672181522815830368604993048084492584055528117 \times 1165882340667125990314837655838327081813101225814639260043952099413134433416292453 6139$$

Now, let us return to *Plot*. The result of codebreaker Huang's first attempt to decipher turns out to be indivisible after a series of calculations, which I suppose is an unsuccessful attempt to factor large number N. On her second try, the calculation works, which means that she finds a solution to factor $N = P \times Q$. Of course, it is unlikely that this whole process can be completed on the abacus, which I think is a bit of an exaggeration on the part of the director. Furthermore, this TV series does not make it entirely clear when it mentions the error problem of "Back to the Mainland" cipher system. A cipher system cannot have errors, otherwise decryption would be impossible even with the key. I think the error might be referring to a problem with the construction of the algorithm, such as accidentally finding a composite number for one of (or even both of) P and Q that largely compromises the system's security. If anyone has any idea what this "error" might be in the TV series, please let me know. Lastly, the series refers to John von Neumann as the father of modern cryptography, but this is totally wrong because it is actually Claude E. Shannon. John von Neumann made significant contributions to the invention of modern computers and the proposal of Game Theory, but his work had less to do with cryptography.

In any case, we see that the mathematical principle behind what we consider the most reliable encryption method today is in fact quite straightforward and not enigmatic. It is simply finding a couple large prime numbers and doing some multiplication and division operations. Despite its simplicity, this mathematical principle ensures that cryptography after WWII is almost indecipherable. During the Cold War, both the US and the Soviet Union devoted unprecedented energy to intercepting the other country's intelligence, but there was no significant leak of classified information as a result of encryption failure.

17.3 Summary

As we discussed in the introduction to information theory, information can be used to resolve the uncertainty in a system. The use of known intelligence information to resolve the uncertainty in an intelligence system is decryption. Hence, the ideal cryptography ensures that the intelligence system's uncertainty cannot be resolved regardless of how many ciphertexts the enemy intercepts. In order to achieve this goal, we have to create ciphertexts that not only are independent of one another but also appear to be in a completely random order. After the birth of information theory, scientists devised remarkable cryptography methods, and public key cryptography is the most commonly used encryption method at present.

Chapter 18

Not all that glitters is gold: Search engine's anti-SPAM problem and search result authoritativeness question

When using search engines, we all hope to obtain useful and authoritative information rather than webpages that only seem to be relevant. Unfortunately, there is no perfect, noise less search result. Sometimes the noise is artificial, mainly SPAM that targets search engine page rankings; other noise could be the result of user activities on the internet, such as large amounts of inaccurate information created by users and non serious compilers. Although noise cannot be avoided completely, good search engines try to eliminate it to the best of their abilities in order to provide the most relevant and accurate search results to the users.

In this chapter, we will discuss how a search engine filters SPAM and provides the most authoritative results in two sections.

18.1 Search engine anti-SPAM

Since search engines came into existence, SPAM pages have been targeting the search engine's result rankings. As a result of commercial SPAM pages, users do not always find high-quality, relevant pages among the high-ranking pages on a search engine. In other words, not all that glitters is gold.

While the ways to spam are numerous, their shared goal is to boost page ranking through illegitimate means. In the early days, a common SPAM method was repeating keywords. For example, for a website that sells digital cameras, it would repeatedly enumerate all kinds of digital camera brands, such as Nikon, Canon, and Kodak. In order to hide the annoyingly repetitive keywords from the reader's eye, smart spammers would often use tiny fonts and text colors that are identical to the background. However, this kind of spamming is easily detected and corrected by the search engine.

After PageRank algorithm was created, spammers realized that the more a page is cited, the higher its ranking. Hence came forth the link trading business. For example, someone might create hundreds of thousands of websites without any content except for their clients' website links. This approach is quite clever compared to the repeating keywords method because these clients' websites themselves do not have any incriminating content, thus hiding the spamming behavior behind the curtains. However, this trick can still be seen through. Because those who help others boost their page rankings need to sell links in large quantities in order to keep up their business, their identities are easily discovered. (This is analogous to producing counterfeit money: once large amounts of counterfeit money is in circulation, it becomes easy to locate the source.) Later, all sorts of ways to spam came along, and I will not enumerate all of them here.

In 2002, my first project after joining Google was to eliminate search engine SPAM. Search engine SPAM had been a severe problem around that time. Back then, anti-SPAM was new, and the spammers had no idea that we were out to catch them. After collective efforts of several months, my colleagues and I successfully eliminated over half of the spammers, and we subsequently caught most of the remaining ones. (Of course, the efficiency of catching spammers would not be so high in subsequent years.) The spammers were caught by surprise. While some of the websites decided to "repent" and forgo their old ways, many of them continued to spam using updated tricks. We had anticipated this and prepared to catch them again. Hence, anti-SPAM became a long-term cat-and-mouse game. Although there is no way to solve the SPAM problem once and for all at present, Google can discover and eliminate known cheating behavior within a certain timeframe to keep the number of SPAM websites to a minimum.

The Chinese speak of the realm of principle and the realm of technique when it comes to accomplishing things. The same applies to search engine anti-SPAM work. In the realm of technique, we see an instance of cheating, analyze it, then eliminate it. This approach solves the problem without demanding too much thinking. However, it is hard to distill a general pattern from individual cases despite a large amount of grunt work. Many search engine companies that advocate for "artificial" prefer this approach. In the realm of principle, we find the motivation and the nature of SPAM from a specific example, then solve the problem from its root.

We realize that communication model still applies to search engine anti-SPAM. In communication, there are generally two ways to solve the noise interference problem:

1. Starting from the source, strengthen the anti-interference capability of the communication (coding) itself.

2. Targeting transmission, filter out interfering noise to restore original information.

Fundamentally, search engine SPAM is noise added to the (search) ranking information, so the first thing to do is strengthening ranking algorithm's anti-noise capability. Then, we need to restore the original, true rankings like noise elimination in signal processing. Readers with experience in information theory and signal processing might be familiar with this fact: if you make a call with your cellphone in a car with a loud engine, the other person probably cannot hear you too clearly; however, if you know the frequency of the car engine, you can add a same-frequency, opposite-amplitude signal to the engine noise to cancel it out, and the call receiver would not hear the car's noise at all. In fact, many high-end cellphones already have this noise detection and cancelation function. The process of noise elimination can be summarized in Figure 18.1:

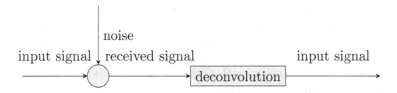

FIGURE 18.1: Noise elimination process in communication.

In the above diagram, the original signal is mixed with noise, which is equivalent to two signals' convolution mathematically. The noise cancellation process is a deconvolution process. This is not a difficult problem in signal processing. First, the car engine has a set frequency. Second, the noise at this frequency appears repeatedly, so we only need to collect several seconds of signal to process it. Broadly speaking, as long as the noise is not completely random and uncorrelated, we can detect and eliminate it. (In fact, completely random and uncorrelated white Gaussian noise is most difficult to eliminate.)

What spammers do in search engine SPAM is similar to adding noise to cellphone signal, which makes the search ranking result chaotic. However, this kind of artificially added noise is not hard to eliminate, since the cheating method cannot be random (otherwise, it would not be able to boost rankings). Moreover, spammers cannot adopt a new approach every day, which means that the cheating method is time-correlated. Therefore, people working on search engine ranking algorithm can detect spammers and restore original rankings after collecting cheating information over a period of time. Of course, this process takes time, just as collecting car engine noise needs time. During this timeframe, the spammer might already be reaping some benefits. Hence, some people think this kind of so-called optimization (which is actually cheating) is effective because they see a short-time improvement for their websites' rankings. Nonetheless, the "optimized" ranking will fall back before long. This does not mean that the search engine used to be generous and is now stricter, but demonstrates that catching spammers and detecting SPAM websites require a certain amount of time.

Motivationally speaking, spammers only want to boost the rankings of their websites in order to obtain commercial benefits. Those who help others cheat (they call themselves Search Engine Optimizer, SEO) also reap benefits. Now, we can prevent spamming behavior by targeting their motives. Specifically, using a set of strong "anti-interference" search algorithm for business-related searches is analogous to using an anti-interference microphone in a noisy environment. For information searches, using "sensitive" algorithm is comparable to using a sensitive microphone for small sounds in a quiet environment. Those websites selling links all have large amounts of out-links, but these out-links have vastly different characteristics compared to out-links on non-SPAM websites (although they might not realize this themselves). The number of out-links on a website linking to other websites can be a vector, and it is a characteristic specific to this website. Since it is a vector, we can calculate the cosine distance (cosine theorem comes in handy again!). We notice that the cosine distance between some websites' out-link vectors is very close to 1. Generally speaking, these websites are typically built by an individual with the singular purpose of selling links. After discovering this pattern, we improve the PageRank algorithm so that bought links become ineffective.

Another anti-SPAM tool is graph theory. In a graph, if several nodes are all connected to one another, they are called a clique. SPAM websites usually need to be linked to each other in order to boost rankings. Thus, they form a clique in the large graph of the internet. There are specific methods to detect cliques in graph theory that can be applied to anti-SPAM directly. Here, we see the efficiency of mathematical principles once again. In terms of techniques, there are many methods. For instance, we can target the JavaScript of SPAM jump page and analyze the corresponding JavaScript content.*

Lastly, there are several points to emphasize. First, Google's anti-SPAM and ranking-restoration processes are completely automated (without any personal preferences), just like the automated process of cellphone noise cancellation. If a website wanted to have high ranking in the long term, it has to have good content and avoid involving itself with any SPAM websites. Second, most people who help others cheat only target search engine algorithms with the largest market shares because spamming also has costs; SPAM targeting a search engine with less than 5% of the market share would not be economically cost-effective. Therefore, having a few instances of spamming for a small search engine does not necessarily indicate advanced anti-SPAM technology, because there are very few people who would cheat there.

In recent years, with constant investment into mainstream search engine anti-SPAM, the cost of spamming is rising higher in most countries that it exceeds the cost of advertising on search engines. Now, most businesses that

---

*Many SPAM websites' landing pages have high-quality content, but they embed a JavaScript that jumps to another commercial website. Hence, after the user enters this page, the landing page only flashes before it jumps to the SPAM page. Search engine crawlers will construct content index according to the high-quality content of the landing page. When the user looks up information, these landing pages are ranked highly because of their content, but the user is taken to the SPAM page with irrelevant information.

hope to boost their websites' rankings choose purchasing search advertisements to increase traffic rather than turning to SPAM. Some respectable websites have also drawn a line between themselves and SPAM websites. However, the opposite trend is true in China: some websites, including government websites, sell links for economic gains. This gave birth to a "gray industry" of brokers who buy and sell links. Of course, the fox leaves some traces when it runs across the field, which helps the hunter track it down.

Web search anti-SPAM is a long-term commitment for a search engine company. The nature of SPAM is adding noise to page ranking signal, so the key to anti-SPAM is noise elimination. Following this logic, we can fundamentally strengthen the search algorithm's anti-SPAM capability to achieve the best results with the least amount of work. If we only targeted spamming behavior case by case with stop-gap measures, it would be easy to let the spammers lead us by the nose.

18.2   Authoritativeness of search results

Users generally use search engine for two purposes. The first one is navigation, which is locating the desired website through search engine. This problem has been solved very well today. The other purpose is finding information. The search engines today can provide a lot of information for almost any searches, but the reliability of the information, especially when the user's question requires professional answers (medical questions for example), cannot always be guaranteed. As the scale of the internet grows, all kinds of inaccurate information multiplies. How to discern the most authoritative information among many sources has become a challenge to search engine companies in recent years. This is one of the problems I tried to solve after joining Google for a second time in 2012.

Some readers may ask, doesn't Google have means like PageRank to assess the content quality of webpages? Can these methods not solve the authoritativeness problem of search results? The first thing we need to point out is that PageRank and other page quality assessment methods have a difficult time evaluating the authoritativeness of search results. For example, many media outlets aim to entertain the public rather than to provide accurate information. Although they may be well known with high PageRank and well written articles, their content is not necessarily authoritative because their primary purpose is entertainment (such as the American website, people.com, and the Chinese forum, tianya).

Moreover, there are often contradictory answers on the internet to the same question. For example, there are almost a hundred answers to the question of Obama's birthplace. His political opponents say that he was born in Kenya, though of course the official answer is Hawaii. How do we know which answer is the most authoritative? Most people with some commonsense would know better than to trust the political opponents' accusations, but how could a computer tell?

When it comes to highly professional questions, people often see contradictory or ambiguous answers. For this kind of questions, even famous news websites cannot guarantee authoritative answers. For example, CNN addressed the question of whether cellphone radiation causes cancer using the research result from the World Health Organization (WHO).* Although it appears to be very reliable, you might think otherwise if you found the actual WHO research report referenced in the article. The report roughly says: "cellphone radiation is classified as Group B, which may be carcinogenic... but up to this date, there is no conclusive evidence that cellphone use will cause cancer." The WHO report only says that cellphone use may cause cancer without any mention of its likelihood. Because there is no definitive conclusion, the research report essentially takes a neutral stance. However, although the CNN article does not claim to have the final say, its tone implies that using cellphones could increase the risk of cancer. Hence, if you were to search the question, "does using cellphone cause cancer?", CNN might not be the most authoritative site. In fact, authoritative medical websites such as the American Cancer Society (cancer.org) and the Mayo Clinic answer this question very cautiously. They do not give a definitive conclusion, but provide a lot of information for the reader to see for themselves.

As the two examples above illustrate, neither "Obama was born in Kenya" nor "cellphone use can cause cancer" can be evaluated as incorrect from the perspective of relevance and search quality, but these statements lack authoritativeness.

Now, how do we assess authoritativeness? To demonstrate this point, we need to introduce the concept of "mention." For instance, a news article might have the following description:

> According to the research findings of WHO, smoking is bad for health.

Or

> A professor from Johns Hopkins points out that second-hand smoke is equally detrimental to health.

When discussing the topic "the harm of smoking," we "mention" "WHO" and "Johns Hopkins." If these two organizations are often "mentioned" in news, scholarly articles, and other webpages on this topic, then we have the reason to believe that these two organizations are authoritative when it comes to "the harm of smoking."

We need to point out that "mention" information is not as apparent upon first glance or easily obtainable like superlinks between pages. It hides in natural

---

*Cell phone use can increase possible cancer risk, http://www.cnn.com/2011/HEALTH/05/31/who.cell.phones/

sentences in articles and needs to be analyzed with natural language processing methods. Even with a good algorithm, the amount of computation required is very large.

Another challenge in evaluating website or webpage authoritativeness is that different from regular page quality (such as PageRank), authoritativeness is search topic-related. For example, WHO, Mayo Clinic, and the American Cancer Society are highly authoritative when it comes to the medical field, but not for finance. On the other hand, CNN may not be authoritative for medical advice, but it is a relatively authoritative source when it comes to public opinion, political views, and general news. Authoritativeness is related to the search keyword. This characteristic dictates an enormous storage. For example, suppose we have M webpages and N keywords, then we need to compute and store $O(M \cdot N)$ results. On the other hand, computing general webpage quality is much simpler, which only requires calculating and storing M results. Hence, only with today's cloud computing and big data technologies can calculating authoritativeness become a possibility. The steps to computing authoritativeness can be summarized as follows:

1. Perform syntactic analysis for every sentence in every page's main text, including title (we will introduce the methods for syntactic analysis in more detail later), then find all phrases related to the topic (e.g., "the harm of smoking") and descriptions of information source (e.g., WHO, Mayo Clinic). Now we have obtained what we have been calling "mention" information. It is worth pointing out that the amount of computation required to analyze every sentence in billions of pages is enormous. Fortunately, Google's syntactic analyzer is fast enough with many available servers, making this challenging feat a reality.

2. Using mutual information, find the correlation between topic phrases and information sources. We have mentioned this method previously.

3. Cluster topic phrases. Although many phrases appear to be different, they may have similar meanings, such as "the harm of smoking," "does smoking cause cancer," "the harm of cigarettes," "the harm of tobacco," and so forth. After clustering these phrases, we have some search topics. As for the clustering method, we can use the matrix operations introduced previously.

4. Lastly, we need to cluster the pages in a website according to the subdomain or sub-directory. Why is this step necessary? This is because for an authoritative website, not all sub-domains are necessarily authoritative. For example, Johns Hopkins University's website may have many medically irrelevant sub-domains, such as medical school students extracurricular activities. Therefore, authoritativeness evaluation has to be constructed at the level of sub-domain or sub-directory.

After completing the four steps above, we can obtain an associative matrix of information source (websites) authoritativeness on different topics. Of course, when computing this associative matrix, we can emulate the approach for computing PageRank and assign greater weight to "mentions" in highly authoritative websites. Then, through iterative algorithms, we can derive a converged associative matrix of authoritativeness. With this matrix, we can promote search results from authoritative information sources to provide more reliable information for users.

When computing authoritativeness, we employ methods from three other chapters, including syntactic analysis, mutual information, and phrase clustering, all of which are supported by mathematical principles. Hence, it is not an overstatement to say that evaluating search result authoritativeness is completely built upon the foundation of mathematical models.

18.3   Summary

Noise is existent in any communication system. Good communication systems need to be able to filter out noise and restore the original signal. Search engines are a special communication system. While noise is inevitable, anti-SPAM and authoritativeness assessments are the process of noise elimination that relies on mathematical methods.

Chapter 19

Discussion on the importance of mathematical models

You may have already noticed that regardless of the problem or questions at hand, we always search for an accurate mathematical model to fit them. In order to clarify the importance of mathematical models, I spent two hours to introduce the mathematical model of search engines when I taught the class "The Fundamentals of Internet Search" at Google China in 2006. Later on, when I joined Tencent in 2010, I discussed the same materials in my first lectures to its engineers.

Among all the astronomers, including Copernicus, Galileo, and Newton, I admire Claudius Ptolemy (90AD - 168AD) the most, who developed the geocentric model.

Astronomy originated in ancient Egypt and Mesopotamia. In Egypt, the Nile River flooded once every year, the lower reaches of the river were easy to irrigate and the soil was extremely fertile. From there was born the earliest human agricultural society. After each flood, the Egyptians would till and sow the recently flooded Earth. This would ensure a plentiful harvest. This farming method persisted until the 1960s, when the Aswan Dam was completed. Afterwards, the lower reaches of the Nile River never again had the floodwaters irrigate the Earth. (Due to this, a few thousand years of continued Egyptian agricultural tradition has mostly disappeared.)

In order to accurately predict when the floodwaters would come and recede, Egyptians developed the science of astronomy 6,000 years ago. Contrary to how we may think today, ancient Egyptians determined their time-keeping and calendar system based on the positions of the Sirius star and the Sun together. The ancient Egyptian calendar does not have a leap year and so the Egyptians established a link between the heliacal rising of the Sirius and that of the Sun. Egyptians noticed that after $365 \times 4 + 1 = 1461$ years, the Sun and the Sirius rise together at the same position. In fact, using the rising positions of both the Sun and the Sirius Star, it is more accurate to measure the time of a year than only using the rising of the Sun. The ancient Egyptians were able to predict where and when the flood waters would come, and where and when they should plant crops.

In Mesopotamia, another center of early human civilization, the Babylonians had a more advanced understanding of astronomy. Their calendar had months and four seasons. Moreover, they observed that the five planets

FIGURE 19.1: The trajectory of Venus as seen from Earth.

(Mercury, Venus, Mars, Jupiter, Saturn; the naked eye cannot see Uranus or Neptune) did not just simply revolve around the Earth, but moved in a wave-like manner. In Greek, the word "planet" connotes a planet that drifts. They also observed that the planets move faster at the perihelion than at the aphelion. Figure 19.1 shows the trajectory of Venus as seen from Earth. (Those who have read the *Da Vinci Code* will know, Venus draws a five-pointed star every four years.)

However, credit for the calculation of celestial trajectories and for the creation of the geocentric model goes to Claudius Ptolemy, who lived nearly 2000 years ago. Although we know today that Ptolemy made the simple mistake that the Sun revolves around the Earth - those who truly understand Ptolemy's contribution will be in awe of what he accomplished. In the past several decades, due to political reasons in mainland China, Ptolemy had always been criticized as being representative of erroneous theory. Thus, Chinese people were largely unaware of Ptolemy's unparalleled contribution to astronomy. I only realized his greatness through reading some books about the history of science when I was in the United States. As a mathematician and an astronomer, Ptolemy made many inventions and contributions to the field, and any one of them would be enough to secure him an important position in the history of science. Ptolemy invented geographical coordinates (which we still use today), defined the equator, the meridian, longitude and latitude (today's maps are still drawn like this). He also proposed the heliocentric orbit and developed the radian system (although students might still feel that the system is a bit abstract when learning it in middle school geography class).

Of course, Ptolemy's most controversial contribution to astronomy is the perfection of his geocentric model. Even though we now know that the Earth orbits the Sun, at that time, from what humans could observe, it was easy to conclude that the Earth is the center of the universe. Zhang Heng, a famous

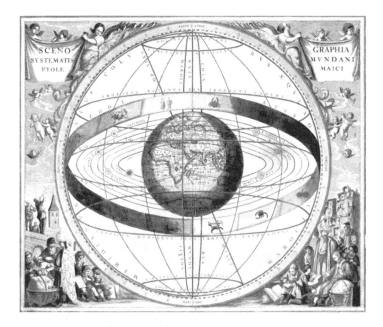

FIGURE 19.2: Ptolemy's geocentric model.

astronomer in ancient China, put forth a theory that is essentially a geocentric model. However, Zhang Heng did not carry out a qualitative description. From Figures 19.2 and 19.3, we can see that the two figures are very similar. However, because Zhang Heng is considered the pride of China, he is always showered with positive propaganda in history books while Ptolemy has become the representative for idealism. In fact, Ptolemy's astronomy is on par with Euclidean geometry and Newtonian physics.

Of course, observing from the Earth, we see irregular planetary motions. However, Ptolemy's genius was in developing a method that entailed using 40-60 small circles inscribed within a larger circle to calculate the planet trajectories with high precision, as shown in Figure 19.4. Ptolemy carried on the legacy of some of Pythagoras' ideas. He believed that the circle was the most perfect shape and thus used the circle to describe the movement of planets.

The high precision of Ptolemy's model amazed later generations of scientists. Today, even with the help of computers, it is still a big challenge to solve equations of 40 inscribed circles. Every time I consider this, I admire Ptolemy from the bottom of my heart. The Julian Calendar matches Ptolemy's calculations, i.e., every year has 365 days, and every four years we have a leap year which adds an extra day. Over the span of 1,500 years, people relied on Ptolemy's calculations for the agricultural calendar. However, 1500 years later, the accumulated error of Ptolemy's proposed path of the Sun is equal to about 10 extra days. Due to this error of 10 days, European farmers were

FIGURE 19.3: Zhang Heng's celestial globe, which is very similar to the geocentric model.

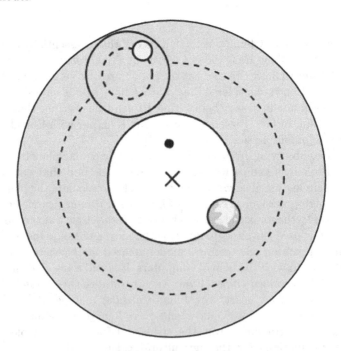

FIGURE 19.4: Ptolemy's geocentric model, using smaller circles encapsulated by larger circles.

off by almost one whole solar period, which had a huge impact on agriculture. In 1582, Pope Gregory XIII took off 10 days from the Julian Calendar. Furthermore, the leap year following every century was changed into a regular year, but the leap year after every 400 years remained a leap year. This is the solar calendar that we still use today, and is almost completely free of error. In order to commemorate Pope Gregory XIII, this calendar is now called the Gregorian Calendar.

Even though the makeshift calendar designed by Pope Gregory XIII that subtracts 3 leap years every 400 years is more accurate, it is hard to learn from his example since he did not propose this solution from any theoretical basis. Although the Gregorian Calendar reflects the movement cycle of the Earth accurately, it does not provide much help for the other planets' orbital patterns. Furthermore, in order to correct the errors of the geocentric model, we can no longer simply inscribe more circles in Ptolemy's 40-circle model. We need to take another step in exploring the truth. Nicolaus Copernicus from Poland discovered that if we use the Sun as the center when describing planetary orbit, we only need 8 to 10 circles to calculate a planet's trajectory. Thus, he put forth a heliocentric model. Regrettably, Copernicus' correct assumptions did not yield him better results than Ptolemy. Compared with Ptolemy's model, Copernicus' errors were much larger. For fear of angering the Catholic Church, Copernicus did not dare to publish his theory until right before his death. At first, the Church did not realize the revolutionary nature of this theory and did not prohibit it. However, when it became clear that this piece of scholarship had the potential to dispute the idea that God had created the world, the Church began to forbid it. Furthermore, the inaccuracy of Copernicus' heliocentric model was another important reason that the Church and the people of that time believed Copernicus' theory to be evil. Therefore, in order for people to be convinced of the heliocentric model, it needs to describe planetary movement more accurately.

The one to finish this mission was Johannes Kepler. Among all the astronomers, Kepler lacked talent, having made careless errors from time to time. However, he had two advantages that other people did not have. First, he inherited lots of accurate observation data from his teacher, Tycho Brahe. Second, he had good luck. Kepler discovered that, in reality, the planets revolve around the Sun in an elliptical orbit (Figure 19.5). This way, there is no need to inscribe many small circles in a larger circle, because one ellipse would suffice to clearly describe the pattern of planetary movement. From this, Kepler proposed three laws that are all very simple.* Kepler's knowledge and expertise were not quite enough to explain why the planets' orbits are elliptical.

---

*Kepler's First Law: All planets move in elliptical orbits, with the Sun at one focus. Kepler's Second Law: A line that connects a planet to the Sun sweeps out equal areas in equal times. Kepler's Third Law: The square of the period of any planet is proportional to the cube of the semi-major axis of its orbit.

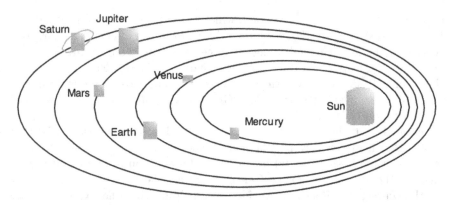

FIGURE 19.5: Kepler's model of planetary orbits.

The glorious and momentous task of explaining why planetary orbit is elliptical was finally completed by the great scientist, Isaac Newton, whose law of gravity demonstrated this very clearly.

The story did not end there. Many years later, more issues arose. In 1821, French astronomer, Alexis Bouvard, discovered that Uranus' actual trajectory is not consistent with the elliptical model. Of course, the lazy approach would be to continue using small circles inscribed in a larger circle to correct this mistake, but many serious scientists searched very hard for the real cause of this error. During the period of 1861 to 1862, a British man, John Couch Adams, and a Frenchman, Urbian le Verrier, independently discovered that Neptune attracted Uranus from its own orbit.*

At the end of my lecture, I presented the engineers from Google China and Tencent with the following concluding points:

1. An accurate mathematical model should be simple in form. (Ptolemy's model was obviously too complicated.)

2. An accurate mathematical model might not be more accurate than a carefully crafted erroneous model at first. However, if we think that the general direction is correct, then we must stick with it. (At the beginning, the heliocentric model appeared no better than the geocentric model.)

3. A large amount of accurate data is important for research and development.

4. Accurate models may be disturbed by noise interference and seem inaccurate. When this happens, we should avoid makeshift fixes

---

*The astronomer, Galileo, actually observed Neptune much earlier in 1612 and 1613, but he mistook it for a fixed star and missed his chance to discover Neptune.

and search for the source of the noise. This may lead to a major discovery.

During the research and development of web searching, the term frequency/ inverse document frequency (TF-IDF) and PageRank that we mentioned in previous chapters can be seen as the "elliptical model" in web searching. They, too, are very easy to understand.

Chapter 20

Don't put all your eggs in one basket: The principle of maximum entropy

When people talk about investment, they often say, "don't put all your eggs in one basket," meaning that one should reduce the risk factor of an investment. In information processing, this principle also applies. In mathematics, this principle is called the Principle of Maximum Entropy. This is a very interesting and profound topic. However, due to its complexity, we are only able to roughly introduce its elements here.

Web search ranking involves hundreds of kinds of information. When I was working at Tencent, engineers often asked me about the most efficient way to combine them. More generally speaking, we often know all sorts of information with a degree of uncertainty in information processing, which requires us to use a unified model to integrate all the information. How to best integrate this information proves to be a challenge.

Let us look at a simple example of figuring out who is Michael Jordan, the basketball player or the computer science professor of Berkeley. When people search these two keywords in Google, they usually only mean one particular person, not all people of this name. So, it is important to separate search results of Michael Jordan the basketball player, from those of Professor Michael Jordan. Of course, if the article is about sports, then we can infer that it likely refers to the basketball player Michael Jordan; if the article discusses artificial intelligence, then Professor Michael Jordan would be the more probable answer. In the above example, we only need to integrate two different kinds of information, which are topic and keyword context. Although there are many viable approaches, such as independently processing thousands of thousands of topics or taking the weighted average of impact of each kind of information, none of them solves the problem gracefully or accurately. This is analogous to the aforementioned patching method for the planetary motion model in the previous chapter. In most applications, we need to integrate dozens or even hundreds of kinds of information, and this sort of makeshift solution obviously cannot suffice.

20.1   Principle of maximum entropy and maximum entropy model

Mathematically, the most beautiful solution to the problem above is the maximum entropy model. It is comparable to the elliptical model of planetary motion. Maximum entropy sounds very complicated, but its principle is in fact quite simple. We use it every day. In laypeople terms, it means retaining all of the uncertainty to minimize risks. Let us turn to the following real-life example.

Once, I went to AT&T Shannon Labs to give a presentation about maximum entropy models, and I brought a die with me. I asked my audience what the probability of each side facing up was, and everyone said the probabilities would be equal, which means that each side had a probability of 1/6 of facing up. This kind of guess was of course correct. When I asked the audience why they thought so, everyone gave me the same answer again: for this "unknown" die, equal probability is the safest assumption. (You should not assume that it is a crooked die that had been filled with lead.) From the perspective of investment, this is the least risky assumption. From the point of view of information theory, it is keeping the largest uncertainty, which is to say the largest entropy. Then, I told the audience that I had performed a special treatment on the die so that the probability of the four-pip side facing up was 1/3. Then I repeated my question to the audience, asking them what the probability of each side facing up would be under this circumstance. This time, most people thought that subtracting the four-pip side's probability of 1/3, the remaining sides each had a probability of 2/15. In other words, the known factor (1/3 probability for the four-pip face) must be satisfied while the other sides' probabilities are still unknown, so the assumption had to be that they are equal. Note that people did not add any subjective assumptions when making guesses of the probability distributions in these two situations, such as the three-pip side must be the opposite of the four-pip side, etc. (In fact, some dice have the one-pip side on the opposite side of the four-pip side.) This kind of intuition-based guessing is accurate because it coincidentally concurs with the principle of maximum entropy.

According to the principle of maximum entropy, when predicting the probability distribution of a random event, our predictions should satisfy all known factors without making any subjective assumptions about unknown situations. (Making no subjective assumptions is crucial.) Under such circumstances, the probability distribution is most even and risk is minimized. Because the information entropy is the greatest under this distribution, people call this model "maximum entropy model." We often say, "don't put all your eggs in one basket." This is a simple way of expressing the principle of maximum entropy. This is because when we encounter uncertainty, we need to keep all of our options open.

Now, let us return to our example of distinguishing two Michael Jordans. Two kinds of information are known before presenting users the search results:

first, user input to the search engine, i.e., two keywords Michael and Jordan; second, the topics of web pages. We know that there are two topics relevant to Michael Jordan, sports and computer science. Hence, we can construct a maximum entropy model that accommodates for both kinds of known information. Our problem now is whether such a model exists. The famous Hungarian mathematician Imre Csiszr, recipient of the highest annual award in information theory which is the Claude E. Shannon Award, proved that for any given set of non-self-contradictory information, there is one and only one maximum entropy model. In addition, these models have a very simple form—exponential function. The following equation is the maximum entropy model of predicting the likelihood of a web page given keywords and topic. Here, d is the web page (document) to be retrieved, w_1 and w_2 are two keywords, e.g., Michael and Jordan, and s represents a topic.

$$P(d \mid w_1,\, w_2,\, s) = \frac{1}{Z(w_1,\, w_2,\, s)} e^{\lambda_1(w_1, w_2, d) + \lambda_2(s, d)} \tag{20.1}$$

Here, Z is the normalization factor to ensure that the probability adds up to 1.

In the above equation, there are a few parameters λ and Z, which need to be trained with observation data. We will introduce how to train the many parameters of maximum entropy models in the extended reading.

Maximum entropy model is the most beautiful and perfected statistical model in its form, and is widely applied in natural language processing and in finance. In the early days, because the computation required for the maximum entropy model is too large, scientists usually adopted similar models that approximated the maximum entropy model. However, this approximation made the maximum entropy model imperfect. The results that came from these models were a little better than those of the patching methods. Therefore, many scholars who originally advocated for the maximum entropy model gave up on it. The first person to demonstrate the advantage of maximum entropy models in practical information processing application was Adwait Ratnaparkhi, former researcher at Microsoft and Nuance and current Senior Director of Research at Yahoo. Ratnaparkhi studied under Professor Marcus when he was at the University of Pennsylvania. Rather than approximating the maximum entropy model, Ratnaparkhi found several natural language processing problems that are most suitable for maximum entropy models and require manageable amounts of computation, such as speech tagging and syntactic analysis. Ratnaparkhi successfully integrated contextual information, parts of speech (noun, verb, and adjective), and sentence elements using maximum entropy models, producing the most advanced speech tagging system and syntactic analyzer at the time. Ratnaparkhi's paper was refreshing in the field. His speech tagging system remains the most effective one among single-method systems as of today. From Ratnaparkhi's accomplishments, scientists renewed their hope in using maximum entropy models to solve complex text information processing problems.

After the turn of the century, because of improvements in computing speed and training algorithms, many complex problems, including syntactic analysis, language models, and machine translation, can be implemented using maximum entropy models. Compared to some models that simply combine different features, the maximum entropy model is more effective by several percents. To the people and companies who care little for product quality, this improvement may not seem significant for the user. However, if the return of investment increases by even one percent, the resulting profit could be counted in billions. Hence, Wall Street has always preferred the latest technology to maximize trading profits. Because securities (stock, bonds, etc) trading requires taking many complex factors into consideration, many hedge funds have started to use maximum entropy models with successful results.

20.2 Extended reading: Maximum entropy model training

Despite their simple format, maximum entropy models are very complicated in their implementation with huge amounts of computation. Suppose we need to consider 20 different features in our web search ranking, $x_1, x_2, \ldots, x_{20}$, and the page to be ranked is d, then even if all of the features are mutually independent, we will end up with a very "long" maximum entropy model:

$$P(d|x_1, x_2, \ldots, x_{20}) = \frac{1}{Z(x_1, x_2, \ldots, x_{20})} e^{\lambda_1(x_1,d)+\lambda_2(x_2,d)+\cdots+\lambda_{20}(x_{20},d)} \qquad (20.2)$$

Here, the normalization factor is

$$Z(x_1, x_2, \ldots, x_{20}) = \sum_d e^{\lambda_1(x_1,d)+\lambda_2(x_2,d)+\cdots+\lambda_{20}(x_{20},d)} \qquad (20.3)$$

This model involves many parameters λ that need to be obtained through training.

The original maximum entropy model training method is Generalized Iterative Scaling, or GIS. The principle of GIS is not too complicated. It can be roughly summarized into the following steps:

1. Suppose the initial model of the zeroth iteration has evenly distributed probabilities.

2. Use the model obtained from the Nth iteration to estimate the distribution of each kind of information feature in the training data.

3. Repeat step 2, until the model gets converged.

GIS was first proposed by J.N. Darroch and D. Ratcliff in the 1970s. It is a typical Expectation-Maximization, or EM. However, the two original authors did not provide a clear explanation for the physical meaning of this algorithm. This was later clarified by mathematician Csiszar. Hence, when people talk about GIS, they often cite both Darroch and Ratcliff's paper and Csiszar's paper. Every iteration of GIS takes a long time and the algorithm requires many iterations before convergence. Moreover, it is not very stable, and might overflow even on 64-bit computers. In practice, GIS is rarely used. Most people only use it to understand the principle of maximum entropy model algorithms.

In the 1980s, talented twin brothers, Vincent J. Della Pietra and Stephen Della Pietra, at IBM made two improvements to GIS and proposed Improved Iterative Scaling, or IIS. This decreased the training time of maximum entropy model by one to two degrees of magnitude. Thanks to their contribution, maximum entropy model became practical. Even so, only IBM had the computing resources to use maximum entropy models at the time.

Hence, the vast amount of computation required for maximum entropy models remained an obstacle. When I was pursuing my PhD at Johns Hopkins University, I thought a lot about how to minimize a maximum entropy model's amount of computation. For several months, I derived equations with a pen and sheets of paper, like those mathematicians. Until one day, I told my advisor Dr. Sanjeev Khudanpur that I discovered a kind of mathematical transformation that can reduce the training time of most maximum entropy models by two orders of magnitude compared to IIS. I performed the derivation for him on the blackboard for over an hour, and he did not find any flaws in it. He then went back and thought it over for two days before finally confirming that my algorithm was right. From there, we constructed some very large maximum entropy models. These models were much better than the patching methods. Even after I found the fast training algorithm, it took me three months on 20 Sun workstations (the fastest machines at the time) processing in parallel to train a language model with contextual information, topic information, and grammar information.* This demonstrates the complexity of implementing maximum entropy models, and in particular, the difficulty of efficient implementation. Today, there are still fewer than one hundred people in the world who can implement these algorithms effectively. Readers interested in implementing a maximum entropy model can refer to my paper.[†]

Here, curious readers might want to ask: what did the Della Pietra brothers who first improved the maximum entropy model algorithm end up doing during those following years? After Jelinek left IBM in the early 1990s, the Della Pietra brothers also left academia and showed their talents in finance. Along with many other IBM colleagues who worked on speech recognition, the Della Pietra

---

*With today's MapReduce tools, we could train the model in one day with 1,000 computers processing in parallel.

[†]www.cs.jhu.edu/junwu/publications.html

brothers joined Renaissance Technologies, which had not been a large company then but is now the most successful hedge fund company in the world. We know that there are dozens or even hundreds of factors that determine the rising and falling of stocks, and the maximum entropy method allows us to find a model that can satisfy thousands of different factors at once. At Renaissance Technologies, Della Pietra brothers and other scientists predicted the stock market with advanced mathematical tools like the maximum entropy model and were met with great success. Since its establishment in 1988, the hedge fund's average annual net return is as high as 34%. In other words, if you invested $1 in the hedge fund in 1988, you would have received over $200 in 2008. This performance far surpasses that of Warren Buffet-led holding company Berkshire Hathaway, whose total return over the same period is 16 times. During the 2008 financial crisis, Renaissance Technologies' return was as high as 80%, which demonstrates the power of mathematical models.

20.3   Summary

The maximum entropy model integrates all kinds of information into one unified model. It has many advantages: in terms of form, it is simple and elegant; speaking of effect, it is the only model that can satisfy constraints while ensuring smoothness. Because of these advantages, maximum entropy models are widely applied. However, due to the overwhelming amount of computation, the model's practicality depends on the quality of implementation.

Bibliography

1. Csiszar, I. I-Divergene Geometry of Probability Distributions and Minimization Problems. The Annals of Statistics. Vol. 3, No. 1, pp. 146–158, 1975.

2. Csiszar, I. A Geometric Interpretation of Darroch and Ratcliff's Generalized Iterative Scaling. The Annals of Statistics. Vol. 17, No. 3, pp. 1409–1413, 1989.

3. Della Pietra, S., Della Pietra, V. & Lafferty, J. Inducing Features of Random Fields, IEEE Trans. on Pattern Analysis and Machine Intelligence. Vol. 19, No. 4, pp. 280–393, 1997.

4. Khudanpur, S. & Wu, J. Maximum Entropy Techniques for Exploiting Syntactic, Semantic and Collocational Dependencies in Language Modeling. Computer Speech and Language Vol. 14, No. 5, pp. 355–372, 2000.

5. Wu, J. Maximum entropy language modeling with non-local dependencies, PhD dissertation, www.cs.jhu.edu/~junwu/publications/dissertation.pdf, 2002.

Chapter 21

Mathematical principles of Chinese input method editors

Inputting Asian languages and other non-Roman languages used to be a problem. However, in the past 20 years, China and other Asian countries have made significant progress in input methods, and now this is no longer an obstacle. Consider Chinese input for example. In the past 25 years, input method evolved from natural syllable encoded input, to radical and strokes input, then returned to natural syllable encoded input. Like any kind of development, the return to natural syllable encoded input is not a mere repetition, but rather a substantial improvement.

The speed of Chinese character input depends on the average length of character codings. In other words, it is the number of keystrokes times the amount of time required to locate the keys. Reducing coding length alone does not necessarily increase input speed because the time needed to locate a key may become longer. Improving the efficiency of an input method requires optimizing both factors simultaneously, and this is founded upon mathematical principles. Here we will illustrate, using mathematical tools, how many keystrokes it takes on average to input a Chinese character, and how to design input coding so that the average number of keystrokes approximates the theoretical minimum without out extending the time required to locate a key.

21.1   Input method and coding

Inputting ideographic Chinese characters into a computer is in fact the process of converting an arbitrarily designed information record coding—a Chinese character—into a computerized coding (GB code or UTF-8 code) information. The keyboard is the main input tool, but of course there are other tools such as the handwriting trackpad and the microphone. Generally speaking, basic keys for Chinese character coding on a keyboard are limited to the 26 letters plus the 10 numbers, including a few control keys. Hence, the most direct coding method is making the 26 letters correspond to pinyin. Of course, in order to solve the problem of multiple words having the same pronunciation (and hence the same pinyin), we need the 10 number keys to clear up ambiguities.

Here, Chinese character encoding can be separated into two parts: pinyin encoding (according to standard Mandarin pinyin) and ambiguity-eliminating encoding. Chinese character code length depends on both aspects. Only when both codes are short can Chinese character input speed improve. Early input methods often emphasized the first part and overlooked the second part.

Although QuanPin (Complete Pinyin) input method is the same as standard Mandarin pinyin and hence easy to learn, early pinyin input method used ShuangPin (Simplified Pinyin) instead of QuanPin. In the ShuangPin input method, each consonant and vowel is only represented by one key. The earliest Chinese character input machines, CEC-I and Great Wall 0520, which correspond to the Apple series and the IBM series, respectively, both employ the ShuangPin input method. The phonetic alphabet used in Taiwan is equivalent to ShuangPin. Each company's keyboard character mapping is slightly different. Taking Microsoft's early ShuangPin input method as an example, the corresponding Table 21.1 of vowels and letters on the keyboard is as follows:

TABLE 21.1:   Table of vowels and corresponding keyboard characters.

Vowels	iu	ua	er,uan,van	ue	uai	uo	un,vn	ong,iong
Letters	q	w	r	t	y	o	p	s
Vowels	uang,iang	en	eng	ang	an	ao	ai	ing
Letters	d	f	g	h	j	k	l	;
Vowels	ei	ie	iao	ui, ue	ou	in	iam	
Letters	z	x	c	v	b	n	m	

This input method appears to shorten the code length on the surface, but in fact does not improve input speed. It optimizes one facet at the expense of the overall efficiency. First, the ShuangPin input method increases coding ambiguity: there are only 26 letters on a keyboard, but there are over 50 Chinese consonants and vowels. We can see from the table that many vowels have to share the same letter key. Added ambiguity results in more candidate characters that the user has to choose from, which increases the length of ambiguity-eliminating coding: this necessitates the repeated process of "next page, scan subsequent words." Second, it increases the time required for each keystroke. Because ShuangPin input method is not intuitive, compared to the QuanPin method, it has the additional process of separating a character's pronunciation into consonant and vowel codings. Cognitive science research shows that during unscripted input, this separation step slows down the thought process. Third, ShuangPin has low fault tolerance for phonetically similar pinyin. Front nasal finals "an, en, in" and their corresponding back nasal finals "ang, eng, ing," roll tongue sounds "ch, sh, zh" and corresponding flat tongue sounds (non-roll tongue sounds) have completely dissimilar codings. Except for people living in the Beijing area, the majority of Chinese people often have trouble distinguishing front nasal finals from back nasal finals, and roll tongue sounds from flat tongue sounds. This results in a situation where a user cannot locate the desired character after flipping through several pages only to realize that the consonant or vowel was chosen incorrectly in the beginning. A good input method should not require the user to master the precise phonetics of every character, just as a point-and-shoot camera should not require the user to be an expert in adjusting aperture and shutter speed.

Due to many factors, early pinyin input methods were less than successful, which created an environment for other input methods to flourish. Many different Chinese input methods sprang up over a short period of time. Some reported over a thousand methods, others say over three thousands. In the early 1990s, the number of patents for all sorts of input methods had exceeded one thousand. Some experts commented that the software industry in China was stagnant because everyone devoted energy to making input methods. Most methods, except for a minority that improved the pinyin input method, most of them directly coded about 6,300 common characters from the Chinese character library (only level 2 national standard characters were considered then) with 26 letters and 10 number keys. We know that even if only the 26 number keys were used, three-key combinations can represent $26^3 \approx 1,7000$ characters. Hence, all of these input methods asserted that they could input a character with only two or three keys: common words only required two keys, and three keys would suffice for uncommon ones. In fact, this is easily achieved and requires little expertise. However, it is almost impossible to remember these complicated codings, so the challenge was associating codings with characters' radicals, strokes, or pronunciations to facilitate memorization. Of course, every coding method claimed to be wiser and faster than all the other methods. Because all coding algorithms of these input methods are at the same level from the perspective of information theory, there is no real superiority or inferiority among them to speak of. However, in order to prove that one method is faster than all the others, people continued on the wrong path and chased after the least amount of keystrokes. The most direct method is to encode word phrases, but this makes memorizing the codings even more difficult for the users and only those who perform these input methods could remember. At this point, the competition was not technical but commercial. In the end, Yongmin Wang's Wubi input method won out temporarily, but this was the result of his better marketing skills (most of the other developers were only book smart) rather than a superior coding method. Today, even Wubi input method has little marketshare. This generation of products is now completely outdated.

The problem with this generation of input methods was that although they decreased the number of keystrokes for each character, they overlooked the time required to locate each key. Requiring the average user to memorize all of the character codings in an input method is not realistic; that would be a harder feat than memorizing 6,000 GRE words. To disassemble a character into codings according to rules when using these input methods was necessary, and the process of finding the coding combination of a character was rather long. This was highly disruptive to one's thought process when typing unscripted. In the beginning of this book, I have emphasized that the major purposes of using language and words are to communicate with others and to express ideas and thoughts, as well as to memorize and to record information. If an input method disrupts the train of thought, then it is incompatible with people's natural behavior. Cognitive science has shown that people are bad at multitasking. In the past, we have conducted many user testings in our voice recognition research and found that when typing unscripted using complex coding input methods, the input speed is only a half to a quarter of the speed when typing according to a script. Therefore, despite a low average number of keystrokes for each character, these input methods slowed down the typing speed so that the overall speed did not increase. It was only natural that the vast number of Chinese computer users did not opt for this type of input methods.

In the end, users chose the pinyin input method, specifically the QuanPin input method that has longer codings for each character. Although it appears that more keystrokes are required for each character's input, it has three advantages so that the speed is by no means slow. First, it does not require special knowledge to use. Second, input is intuitive and does not interrupt the train of thought. In other words, the time required to locate each key is very short. Third, because of the relatively long coding, the information redundancy improves the fault tolerance. For example, for users who have trouble telling front nasal finals "an, en, in" from back nasal finals "ang, eng, ing," if they mistakenly input a back nasal final "zhang" for the word "zhan," they would see the desired word halfway through the inputting process and stop there, thus avoiding the low fault tolerance problem with ShuangPin. Therefore, the only problem left for the pinyin input method is eliminating ambiguity for a pinyin that corresponds to multiple words. If this problem is solved, the pinyin input method can achieve a similar number of keystrokes compared to the word decomposition methods. Hence, ambiguity elimination is the primary project for all kinds of pinyin input methods. Next, let us analyze the least average number of keystrokes for inputting a Chinese character.

21.2   How many keystrokes to type a Chinese character? Discussion on Shannon's First Theorem

Theoretically, how fast can one input Chinese characters? Here, we need to use Shannon's First Theorem in information theory.

The GB2312 simplified Chinese character set has about 6,700 common characters in total. Without consideration of the distribution of the frequency of Chinese characters, using 26 keyboard letter keys to encode characters, two-letter combinations can only encode 676 whereas over 6,700 characters require three-letter combinations. Thus, the coding length is three. Of course, intelligent readers will notice that if we were to use shorter codings for common words and longer codings for uncommon words, the average coding length can be reduced. Suppose the relative frequency of each character's appearance is

$$p_1, p_2, p_3, \cdots, p_{6700} \tag{21.1}$$

and their coding lengths are

$$L_1, L_2, L_3, \cdots, L_{6700}. \tag{21.2}$$

Then, the average coding length is

$$p_1 \cdot L_1 + p_2 \cdot L_2 + p_3 \cdot L_3 + \cdots + p_{6700} \cdot L_{6700} \tag{21.3}$$

According to Shannon's First Theorem, for any piece of information, its coding length cannot be smaller than its information entropy. Hence, the average coding length's minimum is the characters' information entropy. No input method can outdo this lower bound set by information entropy. We need to point out here that even if we were to expand the character set from national standard GB2312 to a larger set GBK, the average coding length would not increase significantly because uncommon characters from the latter set have extremely small frequencies. Therefore, our discussions in this book will use the GB2312 character set.

Now, recall the information entropy of Chinese characters (see Chapter 6 "Quantifying Information")

$$H = -p_1 \cdot log\, p_1 - p_2 \cdot log\, p_2 - \cdots - p_{6700} \cdot log\, p_{6700} \tag{21.4}$$

If we took every character into account regardless of contextual relevance, then a rough estimate of its value is less than 10 bits. Of course, the chosen corpus from which we make estimations also matters. Suppose the input method can only use 26 letter keys for input, then each letter can represent $log\, 26 \approx 4.7$ bits of information, which is to say that inputting one character requires on average $10/4.7 \approx 2.1$ keystrokes. Intelligent readers may have already noticed that grouping characters into words and computing information entropy in units of words will decrease the average information entropy for each character. This way, the average number of keystrokes for inputting a character can be reduced by a decimal. Not considering contextual relevance, using words as the unit of computation, a character's information entropy is about 8 bits. In other words, inputting a character in units of words only needs $8/4.7 \approx 1.7$ keystrokes on average. This is the fundamental reason that all of the input methods at present are word-based. If we were to take contextual relevance into consideration, constructing a word-based statistical language model (see Chapter 3 "Statistical Language Model"), then we can further reduce the information entropy of a character to about 6 bits. Now, inputting one character only requires $6/4.7 \approx 1.3$ keystrokes. If an input method can achieve this, it would be much faster to input Chinese than English.

However, no input method can achieve such efficiency. There are two main reasons. First, in order to approach the limit given by information theory, we need to construct special codings for words according to word frequencies. As we have discussed in the previous section, overly specialized codings aim to expedite but in fact slow the input process down. Moreover, it is very difficult to install large language models on personal computers. Therefore, this kind of coding is theoretically effective but actually impractical.

Now, let us consider the average number of keystrokes needed to input a character using QuanPin pinyin input method. The average length of QuanPin pinyin is 2.98. As long as pinyin-based input methods can solve the ambiguity problem using contextual information, the average number of keystrokes should be within 3, and inputting 100 characters in a minute is completely possible. With greater use of contextual relevance, the character in a sentence can be cued out before its pinyin is completely typed out. Hence, QuanPin's average keystrokes number should be smaller than 3.

The next task is employing contextual information. Pinyin input methods (ZiGuang for example) from 10 years ago solved this problem by constructing a large vocabulary: phrases became longer, and eventually whole verses of classical Chinese poems became a word that could be typed together. Although this solution helped a little bit, a statistical analysis would reveal that it only solves the problem minimally. In Chinese, despite the large number of long words that could be counted in hundreds of thousands, one-character words and two-character words compose the majority of texts. However, it is these one-character words and two-character words that have the most ambiguity. For example, there are 275 characters that correspond with the pinyin "zhi" (in Google Pinyin Input), and 14 two-character words for "shi-yan." These problems cannot be resolved by increasing the size of the vocabulary. Expanding the vocabulary was more or less an act out of experience and intuition, analogous to the makeshift fixes to the geocentric theory when it failed to explain irregular planetary movements. In order to solve the ambiguity problems, many later input methods, including ones that are popular now, incorporated common words and phrases (such as "I am") into their vocabulary. However, there are 40,000 to 50,000 common one-character and two-character words in the Chinese language, and tens of millions or even hundreds of millions of reasonable combinations of these words that are too many to be included into the corpus altogether. Therefore, enumerating word/character combinations, just like epicycles in Ptolemy's geocentric model, can only approximate the truth but can never reach the truth.

The best way to use contextual information is by employing language models. As long as we admit the efficiency of probability theory, we cannot deny that language models guarantee the best results for phonetic transcription (resolving the ambiguity problem). If the language model does not have a limit in size, then it can reach the maximum input speed set by information theory. However, in making input method products, it is impractical to take up too much of the user's storage space. Hence, all kinds of input methods can only provide the user with a severely compressed language model. Due to the desire to save space or the lack of mastery of pinyin-to-character decoding techniques, some input methods leave out language models altogether. Hence, the input methods today have a long way to go before reaching the theoretical maximum input speed. At present, all companies' input methods (Google, Tencent, and Sogou) are roughly at the same level, and the key to further technical improvement is building an accurate and effective language model. Of course, in order to automatically transform a string of pinyin into Chinese characters with language models, we need the appropriate algorithm. This is our topic for the next section.

21.3   The algorithm of phonetic transcription

Transcribing pinyin into characters and finding the shortest path in navigation share the same algorithm; both are dynamic programming. This may sound like a stretch: what does pinyin input method have to do with navigation?

In fact, we can see Chinese input as a communication problem, and the input method is a converter that transcribes a pinyin string into a character string. One pinyin can correspond to multiple characters. If we connect the corresponding characters of a pinyin string from left to right, it is a directed graph, known as a lattice graph.

Here, $y_1, y_2, y_3, \ldots, y_N$ is a pinyin string that the user inputs; w_{11}, w_{12}, w_{13} are candidate characters for the first pinyin y_1 (we will represent these three candidate characters with variable w_1 in later equations); $w_{21}, w_{22}, w_{23}, w_{24}$ are candidate characters corresponding to y_2, and will be represented with w_2, and so forth. Many sentences can be formed from the first character to the last one, and each sentence corresponds to a path in the lattice graph. Pinyin input method locates the best sentence given its pinyin, according to the context:

$$w_1, w_2, \cdots, w_N = \underset{w \in W}{\mathrm{ArgMax}} P\left(w_1, w_2, \cdots, w_N \mid y_1, y_2, \cdots, y_N\right) \qquad (21.5)$$

In the above graph, this means finding the shortest path from start to finish. In order to do so, we need to first define the distance between two nodes in the graph. Review the simplification method we introduced in the chapter on "Implicit Markov Model":

$$
\begin{aligned}
&= \underset{w \in W}{\mathrm{ArgMax}} P\left(y_1, y_2, \cdots, y_N \mid \overset{\displaystyle w_1, w_2, \cdots, w_N}{w_1, w_2, \cdots, w_N}\right) \cdot P(w_1, w_2, \cdots, w_N) \\
&\approx \underset{w \in W}{\mathrm{ArgMax}} \prod_{i=1}^{N} P(w_i \mid w_{i-1}) \cdot P(y_i \mid w_i)
\end{aligned}
\qquad (21.6)
$$

If we were to negate and take the logarithm of the probability in the above equation simultaneously, which is to define $d(w_{i-1}, w_i) = -\log P(w_i \mid w_{i-1}) \cdot P(y_i \mid w_i)$, the product becomes a sum, and finding the largest probability becomes a problem of locating the shortest path. Thus, we can use the dynamic programming algorithm directly to implement the most important phonetic transcription problem in pinyin input method. Comparing it with finding the shortest distance between two locations in satellite navigation, we can see that the two are analogous. The only difference is that in a navigation graph, the distance between two nodes (cities) is a physical distance, whereas in the

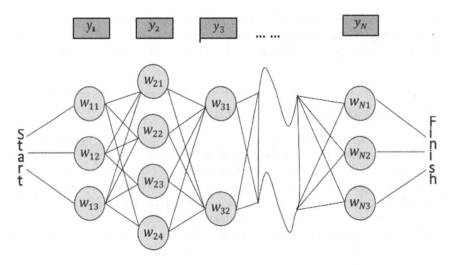

FIGURE 21.1: The lattice graph of pinyin to character transcription decoding.

lattice graph of pinyin string to characters conversion, the distance between two nodes (words) w_{i-1} and w_i is the product of the transfer probability and the generating probability $- \log P(w_i \mid w_{i-1}) \cdot P(y_i \mid w_i)$.

This example of pinyin input method appears to have little connection to the navigation system mentioned in Chapter 12, "Finite state machines and dynamic programming: Navigation in Google Maps", but they are supported by the same mathematical model. The beauty of math lies in the universality of its tools, which can play big roles in many different applications.

21.4   Extended reading: Personalized language models

Suggested background knowledge: probability theory.

Current pinyin input methods for Chinese characters have a long way to go until reaching the limit set by information theory. More and more good input methods will emerge because there is significant room for improvement. Of course, input speed is not the only standard to measure input methods - after the input speed reaches a certain threshold, the user experience may become a more important concern.

Theoretically speaking, as long as the language model is large enough, the average number of keystrokes for pinyin input methods can approach the limit given by information theory. If we put an input method on cloud computing, the implementation is completely within reach. However, this is not realistic on the client (such as a personal computer). Nonetheless, there are advantages to client architecture, such as the ability to construct personalized language models.

The motivation for personalization is that different people usually write about different topics. Due to differences in education background and diction, the level at which one speaks and writes also differs. Hence, different users should have their respective language models.

Our research at Google found this assumption to be correct. People from a variety of places, cultural backgrounds, and educational levels will use different function words, not to mention different content words that convey a topic's meaning. Therefore, if everyone had a personalized language model, the order of candidate words will certainly be better arranged compared to that of a generalized input method when using pinyin to input.

Now, we have two problems to solve. The first is training a personalized language model, and the second is dealing with the relationship between this personalized language model and the generalized language model.

In order to train a specific language model for someone, the best thing to do is to gather enough writings produced by this person, but the writings produced in a lifetime by a single individual are insufficient for training a language model. Training a bigram model with a vocabulary of tens of thousands requires a corpus of tens of millions of words. Even for professional writers or journalists, it is impossible to produce so many articles in a lifetime. Without enough training data, the trained (higher-order) language model is practically useless. Of course, training a unigram model does not require too much data, and some input methods (spontaneously) found an approach out of experience: user dictionaries. This is in fact a small-scale unigram model with a very small number of ordered sets (for example, a user-defined word "ABC" is a triple).

A better solution is to find a large corpus that conforms to the input content and language habits of the user, then train a user-specific language model. The key clearly is how to locate the qualified corpus. Here, we will employ the law of cosines and text classification techniques again. The steps for training a user-specific language model are as follows:

1. Classify articles for training the language model into different categories by topics; for example 1,000 categories C_1, C_2, ..., C_{1000}.

2. For each category, find its feature vector (TF-IDF) X_1, X_2, $X_3, \ldots, X_{1000}$.

3. Using user-inputted articles, compute the feature vector Y of the inputted words.

4. Compute the cosine (distance) between Y and $X_1, X_2, X_3, \ldots, X_{1000}$.

5. Select the articles corresponding to K categories with the smallest distance to Y. Use them as the training data for this user-specific language model.

6. Train the user-specific language model M_1.

Under most circumstances, M_1 works better for this specific user than the generalized model M_0. However, for relatively obscure content, model M_0 would be a lot more effective than M_1. This is because the training data for M_1 is one or two degrees of magnitude smaller than that for M_0, and so the coverage of the linguistic phenomena is much less than that of the former. Therefore, a better approach is to integrate these two models.

We have introduced previously that the best way to integrate different features is to employ a maximum entropy model. Of course, this model is rather complicated with a long training time. If everyone constructed such a model, the cost would be rather high. Hence, we can use a simplified model: linear interpolation model.

Suppose both M_0 and M_1 are bigram models, their conditional probabilities for (w_{i-1}, w_i) are $P_0(w_i \mid w_{i-1})$ and $P_1(w_i \mid w_{i-1})$, respectively. The conditional probability computed from new model M' should be

$$P'(w_i \mid w_{i-1}) = \lambda(w_{i-1}) \cdot P_0(w_i \mid w_{i-1}) + (1 - \lambda(w_{i-1})) \cdot P_1(w_i \mid w_{i-1}).$$

Here, $0 < \lambda(w_{i-1}) < 1$ is an interpolation parameter. Because the information entropy (corresponding to the language model's complexity) is a convex function,* linear combination P''s entropy is smaller than the linear combination of P_0 and P_1's entropies, the new integrated model has less uncertainty and is a better model. In other words, the new model obtained through combining the personalized model and the original generalized model is superior than both.

This kind of linear interpolation model is slightly less effective than the maximum entropy model, but we can still retain 80% of the improvement of the latter. Google Pinyin Input's personalized language models are implemented this way.

21.5   Summary

The process of Chinese character input is communication between people and computers. Effective input methods will consciously or unconsciously follow mathematical models of communication. Of course, in order to produce the most effective input method, we should follow the guidance of information theory.

---

*If a function satisfies the condition $f(tx_1 + (1 - t)x_2) < tf(x_1) + (1 - t)f(x_2)$, then this function is called a convex function.

Chapter 22

Bloom filters

22.1   The principle of Bloom filters

In everyday life and work, including in software development, we often need to determine whether an element is in a set. For example, in word processing softwares, we need to check whether an English word is spelled correctly (whether it is in a known dictionary); the FBI needs to verify whether a suspect's name is on the suspects' list; a web crawler needs to confirm whether a site has already been visited, etc. The most direct solution is to save all of the elements in the set on the computer, and compare each new element against all existing elements in the set. Generally speaking, a set is stored in a hash table on a computer. The main advantages of a hash table are its speed and accuracy, but these come at the expense of storage space. When the set is relatively small, this problem is not apparent. However, when the set is extremely large, the low space efficiency problem of hash tables becomes obvious. For example, public email providers Yahoo, Gmail, and Hotmail have to set up a system to filter out junk emails from spammers. One solution is to record the sender addresses of all spam emails. Because spammers will keep making new email accounts, there are at least billions of spam addresses all around the world that require many servers to store. Using hash tables, we need 1.6 GB of storage for every hundred million email addresses. (The specific implementation is to store 8-bit information fingerprints that correspond to email addresses). Because the storage efficiency of hash tables is only 50%, each email address requires 16 bits of storage. A hundred million email addresses would take up 1.6 GB of storage space. Hence, storing billions of addresses could require hundreds of gigabytes of storage. Other than supercomputers, average servers cannot store such a large number of addresses.

Today, we will introduce a mathematical tool called a Bloom filter. It can solve the same problem using only an eighth to a quarter of the space of a hash table.

Burton Bloom proposed Bloom filters in 1970. It is actually a very long binary vector and a series of random mapping functions. We will use the following example to demonstrate how Bloom filters work.

Suppose we need to store one hundred million email addresses. First, we need to construct a 1.6-billion bits, or 200-million-byte vector, then clear all of the 1.6 billion bits. For every email address X, use 8 different random number generators $(F_1, F_2, \ldots, F_8)$ to generate 8 information fingerprints $(f_1, f_2, \ldots, f_8)$. Then, use a random number generator G to map these 8 information fingerprints onto 8 natural numbers between 1 and 1.6 billion $g_1, g_2, \ldots, g_8$. Now, set all 8 bits to 1. After treating all of the hundred million email addresses with this procedure, we obtain a Bloom filter for these email addresses, as shown in Figure 22.1.

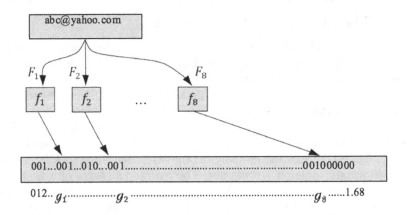

FIGURE 22.1: Bloom filter's mapping method.

Now, let us see how to use this Bloom filter to check whether suspect email address Y is on the blacklist. Using the same 8 random number generators $(F_1, F_2, \ldots, F_8)$, generate 8 information fingerprints for this address $s_1, s_2, \ldots, s_8$, then map these 8 fingerprints to the Bloom filter's 8 bits, $t_1, t_2, \ldots, t_8$. If Y is in the blacklist, then the 8 bits value corresponding to $t_1, t_2, \ldots, t_8$ must be 1. Thus, we can accurately detect any email addresses that are on the blacklist.

This Bloom filter will not omit any suspect email address from the blacklist. However, it also has a drawback. There is a very low probability that an address not on the blacklist will be judged as spam, because the 8 bits corresponding to this "good" email address are "coincidentally" set to 1 (by other addresses) in the Bloom filter. The good thing is that this probability is very low. We call it the false alarm rate. In the example above, the false alarm rate is lower than 1/10000. The theoretical analysis of the false alarm rate will be introduced in the extended reading.

The advantages of Bloom filters are their time and space efficiency, but they also have a certain false alarm rate. A common way to salvage this is to construct a small whitelist to store those email addresses that may be misidentified.

22.2   Extended reading: The false alarm problem of Bloom filters

Suggested background knowledge: probability theory.

In the previous section, we mentioned that one drawback of Bloom filters is that they might misidentify an element that is not in the set as one that is in it. This is called a "false positive" in testing. This probability is very small, but how small is it? Is it negligible?

Estimating the rate of false positive is not difficult. Suppose a Bloom filter has m bits with n elements, and each element corresponds to hash functions of k information fingerprints. Of course, out of these m bits, some are 1 and some are 0. Let us first look at the probability of a bit being 0. For example, if we were to insert an element in this Bloom filter, its first hash function would set a bit in the filter to 1. Hence, the probability of any bit being set to 1 is $\frac{1}{m}$, and the probability of it remaining 0 is $1 - \frac{1}{m}$. For a specific position in the filter, the probability that none of the k hash functions set this element to 1 is $(1 - \frac{1}{m})^k$. If we were to insert a second element in the filter, and this specific position remains 0, then this probability is $(1 - \frac{1}{m})^{2k}$. If n elements were inserted and the position has not been set to 1, this probability is $(1 - \frac{1}{m})^{kn}$. On the other hand, the probability of a bit being set to 1 after inserting n elements is $1 - (1 - \frac{1}{m})^{kn}$.

Suppose n elements have been put in the Bloom filter. For a new element not in the set, because the hash functions of its information fingerprints are randomly generated, the probability of its first hash function having the bit value of 1 is the probability above. For a new element to be misidentified as an element already in the set, all of the bit values corresponding to hash functions have to be one. This probability is

$$(1 - [1 - \frac{1}{m}]^{kn})^k \approx (1 - e^{\frac{kn}{m}})^k \tag{22.1}$$

simplified as

$$P = (1 - e^{-\frac{(\frac{m}{n}ln2)n}{m}})^{(\frac{m}{n}ln2)} \tag{22.2}$$

If n is relatively big, the equation can be approximated as

$$(1 - e^{-k(n+0.5)/(m-1)})^k \approx (1 - e^{-\frac{kn}{m}})^k \tag{22.3}$$

Suppose an element uses 16 bits, where $k = 8$. Then the probability of a false positive is 5/10000. In most applications, this false alarm rate can be tolerated. Table 22.1 shows the false positive rates under different circumstances with

TABLE 22.1:   Table of false alarm rates given different m/n and k.

m/n	k	k=1	k=2	k=3	k=4	k=5	k=6	k=7	k=8
2	1.29	0.393	0.400	0.253	0.160	0.101	0.0638	0.0229	0.0145
3	2.08	0.283	0.237	0.147	0.092	0.0578	0.0464	0.0135	0.00846
4	2.77	0.221	0.155	0.092	0.0561	0.0347	0.0216	0.00819	0.00509
5	3.46	0.181	0.109	0.0609	0.0359	0.0217	0.0133	0.00513	0.00314
6	4.16	0.154	0.0804	0.0423	0.024	0.0141	0.00844	0.00329	0.00199
7	4.85	0.133	0.0618	0.0306	0.0166	0.00943	0.00552	0.00217	0.00129
8	5.55	0.118	0.0489	0.0228	0.0118	0.0065	0.00371	0.00146	0.000852
9	6.24	0.105	0.0397	0.0174	0.00864	0.00459	0.00255	0.001	0.000574
10	6.93	0.0952	0.0329	0.0136	0.00646	0.00332	0.00179	0.000702	0.000394
11	7.62	0.0869	0.0276	0.0108	0.00492	0.00244	0.00128	0.000499	0.000275
12	8.32	0.08	0.0236	0.00875	0.00381	0.00183	0.000935	0.00036	0.000194
13	9.01	0.074	0.0203	0.00718	0.003	0.00139	0.000692	0.000264	0.0014
14	9.7	0.0689	0.0177	0.00596	0.00239	0.00107	0.000519	0.000196	000101
15	10.4	0.0645	0.0156	0.005	0.00193	0.000839	0.000394	0.000147	7.46e-05
16	11.1	0.0606	0.0138	0.00423	0.00158	0.000663	0.000303	0.000112	5.55e-05
17	11.8	0.0571	0.0123	0.00362	0.0013	0.00053	0.000236	8.56e-05	4.17e-05
18	12.5	0.054	0.0111	0.00312	0.00108	0.000427	0.000185	6.63e-05	3.16e-05
19	13.2	0.0513	0.00998	0.0027	0.000905	0.000347	0.000147	5.18e-05	2.42e-05
20	13.9	0.0488	0.00906	0.00236	0.000764	0.000285	0.000117	4.08e-05	1.87e-05
21	14.6	0.0465	0.00825	0.00207	0.000649	0.000235	9.44e-05	3.24e-05	1.46e-05
22	15.2	0.0444	0.00755	0.00183	0.000555	0.000196	7.66e-05	2.59e-05	1.14e-05
23	15.9	0.0425	0.00694	0.00162	0.000478	0.000164	6.26e-05	2.09e-05	1.14e-05
24	16.6	0.0408	0.00639	0.00145	0.000413	0.000138	5.15e-05	1.69e-05	7.16e-06
25	17.3	0.0392	0.00591	0.00129	0.000359	0.000117	4.26e-05	1.38e-05	5.73e-06
26	18	0.0377	0.00548	0.00116	0.000314	9.96e-05	3.55e-05	1.13e-05	
27	18.7	0.0364	0.0051	0.00105	0.000276	8.53e-05	2.97e-05		
28	19.4	0.0351	0.00475	0.000949	0.000243	7.33e-05	2.5e-05		
29	20.1	0.0339	0.00444	0.000862	0.000215	6.33e-05			
30	20.8	0.0328	0.00416	0.000785	0.000191				
31	21.5	0.0317	0.0039	0.000717					
32	22.2	0.0308	0.00367						

Source: http://pages.cs.wisc.edu/~cao/papers/summary-cache/node.html

varying m/n ratios and k (this table was originally provided by Professor Pei Cao at the University of Wisconsin-Madison, who now works at Google).

22.3 Summary

Bloom filters rely on the mathematical principle that the probability of two completely random numbers being identical is very low. Hence, with a very small false alarm rate, Bloom filters can store large amounts of information with relatively little space. A common solution to the false positives problem is to build a small whitelist to store the information that may be misidentified. Bloom filters only involve simple arithmetic, which makes them fast and easy to use.

Chapter 23

Bayesian network: Extension of Markov chain

23.1 Bayesian network

In previous chapters, we often mentioned the Markov chain. It describes a state sequence where the value of each state depends on a certain number of preceding states. For many practical problems, this model is a very rough simplification. In real life, the relationships between different things often cannot be connected with a chain; they could be intersecting and highly intricate. For example, we can see from Figure 23.1 that the relationships between cardiovascular disease and its many contributing factors are too complex to be represented by a chain.

The directed graph above can be seen as a network, each circle representing a state. The curves connecting different states represent their causal relationships. For example, the curve that connects "cardiovascular disease" to "smoking" means that the former may be linked to the latter. If the Markov hypothesis is true in this graph, i.e., a state only depends on its immediately connected states and not on its indirectly connected states, then it is a Bayesian network. However, it is worth noting that the absence of a curve between two states A and B only indicates a lack of direct causal relationship, but this

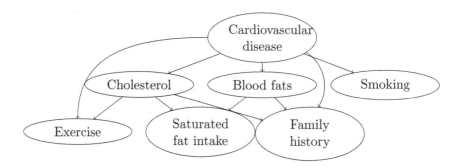

FIGURE 23.1: A simple Bayesian network describing cardiovascular disease and its contributing factors.

does not mean that state A cannot indirectly influence state B through other states. As long as there exists a path in the graph from A to B, these two states are indirectly correlated. All of these (causal) relationships have a quantifiable belief, which means that they can be described by probabilities. In other words, the curves in a Bayesian network can be weighted. The Markov hypothesis ensures that Bayesian networks can be easily computed. We can calculate the probability of someone having cardiovascular disease from a graph like this.

In a network, the probability of any node can be calculated by Bayes' formula. The Bayesian network was thus named. Because every curve in the network has a belief, Bayesian networks are also called belief networks. In the example above, we can simplify the states and suppose that there are only three states: "cardiovascular disease," "high blood fats," and "family history." We can illustrate how probabilities are calculated with Bayes' formula using this simplified example. For the sake of simplicity, suppose each state only has two types of values, "positive" and "negative", as shown in the graph below. The tables represent the conditional probabilities of different values of each state and combined states, respectively. For example, the top table shows that given family history, the probability of high blood fats is 0.4; without family history, this probability is only 0.1.

If we were to calculate the joint probability distribution of "family history," "high blood fats," and "cardiovascular disease," we can use Bayes' formula (FH = family history, HBF = high blood fats, CD = cardiovascular disease):

$$P(FH, HBF, CD) = P(CD \mid FH, HBF) \times P(HBF \mid FH) \times P(FH) \quad (23.1)$$

We can easily calculate the probability using the values in the above table.

Moreover, if we wanted to know how much family history contributes to the probability of cardiovascular disease, we can also use this Bayesian network to calculate.

$$P(\text{has}FH \mid \text{has}CD) = P(\text{has}FH, \text{has}CD)/P(\text{has}CD) \quad (23.2)$$

Here:

$$\begin{aligned} P(\text{has}FH, \text{has}CD) = {} &P(\text{has}FH, \text{has}CD, \text{no}HBF) \\ &+ P(\text{has}FH, \text{has}CD, \text{has}HBF) \end{aligned} \quad (23.3)$$

$$\begin{aligned} P(\text{has}CD) = {} &P(\text{has}FH, \text{has}CD, \text{no}HBF) + P(\text{has}FH, \text{has}CD, \text{has}HBF) \\ &+ P(\text{no}FH, \text{has}CD, \text{no}HBF) + P(\text{no}FH, \text{has}CD, \text{has}HBF) \end{aligned}$$
$$(23.4)$$

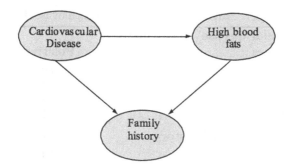

FIGURE 23.2: A simplified Bayesian network.

Each item in the two equations above can be calculated by Equation (23.1). Plugging in the values from Figure 23.2, this probability is

$$P(\text{has}FH, \text{has}CD) = 0.18 \times 0.4 + 0.12 \times 0.4 \tag{23.5}$$

$$= 0.12 \tag{23.6}$$

$$P(\text{has}CD) = 0.12 + 0.18 \times 0.4 + 0.72 \times 0.1 \tag{23.7}$$

$$= 0.12 + 0.144 \tag{23.8}$$

$$= 0.264 \tag{23.9}$$

Plugging this result into Equation (23.2), we see that $P(FH \mid CD) = 45\%$. Hence, although people with a family history of this disease only make up 20% of the population, they count among 45% of all cardiovascular disease patients. In other words, one's likelihood of having cardiovascular disease increases drastically with family history.

In our above calculation of each state in the Bayesian network, only the immediately preceding state was taken into consideration. This is similar to the Markov chain. However, the Bayesian network's structure is more flexible than that of the Markov chain. Without the limitation of a chain-like structure, the Bayesian network can describe correlation relationships more accurately. We could say that the Markov chain is a special case of the Bayesian network, and the latter is a generalization of the former.

In order to use a Bayesian network, we have to first specify its topologic structure and conditional probability distributions of each state given its parent states, as shown in Table 23.1 and Figure 23.2. The processes of obtaining the structure and the parameters are called structure learning and parameter training or learning, respectively. Similar to training a Markov model, we need some known data to train a Bayesian network. For example, to train the network above, we need to know about cardiovascular disease and its relevant information like smoking and family history. Compared to that of a Markov chain, a Bayesian network's learning process is rather complex. Theoretically, it is an NP-complete problem. In other words, it is computationally intractable.

TABLE 23.1:   Statistics of symptoms of cardiovascular diseases.

Cardiovascular disease\family history, high blood fats	Positive	Negative
Positive, Positive	0.9	0.1
Positive, Negative	0.4	0.6
Positive, Positive	0.4	0.6
Positive, Negative	0.1	0.9
High blood fats\family history		
Positive	0.4	0.6
Negative	0.1	0.9
Family history		
	0.2	0.8

However, for certain applications, this process can be simplified and implemented on computers.

It is worth mentioning that Dr. Geoffrey Zweig, former researcher at IBM, and Dr. Jeffery Bilmes, professor at University of Washington, developed a toolkit for generalized Bayesian networks that is freely available for interested researchers.

Bayesian networks have many applications in fields such as image processing, text processing, and decision support systems. In text processing, a Bayesian network can describe the relationships between words with similar meanings. We can use Bayesian networks to locate synonyms and related words, which have direct application in both Google Search and Google Ads.

23.2   Bayesian network application in word classification

We can use a statistics-based model to process articles, summarize concepts, and analyze topics. We call this kind of model a topic model. In previous chapters, we mentioned a type of topic model where we map the eigenvector of an article to the eigenvector of a topic through cosine similarity (distance). This is one kind of statistical topic model. Here, we will introduce another kind of model that uses the Bayesian network, Google's RePhil. Since I cannot go into too much detail, I will use my own language to explain how RePhil works. Those who have listened to presentations or have read PowerPoints about it may notice that my introduction\is slightly different, but the underlying principles are the same.

Before discussing RePhil, let us first review text classification. Suppose there are one hundred million documents, and we can classify them into 10,000 categories by using either singular value decomposition of correlation matrix (of documents and keywords) or cosine distance clustering. Of course, the specific number of categories can be determined by the engineers themselves. Each document can be classified into one category or many categories. Documents in the same category share many keywords. As for how these keywords are generated, there could be many ways. For example, selection by hand. For processing a large amount of data, we have to use automated methods. This way, different documents establish relationships among them through keywords, and these relationships indicate whether certain documents belong to the same category.

If we were to rotate the correlation matrix of documents and keywords by 90 degrees and conduct singular value decomposition, or construct a vector for each word using the document as its dimension then cluster the vectors, then we would obtain a word classification rather than a document classification. We can call each category a concept.

Of course, one concept can include many words, and one word can belong to multiple concepts. Similarly, a document can correspond to many concepts, and vice versa. Now, we can use a Bayesian network to construct relationships among documents, concepts, and keywords.

In Figure 23.3, the document and keywords have direct relations and are both directly related to concepts. They are also indirectly related to each other through topics. Similar to the example of cardiovascular disease and family history, we can find the correlation between a word and a concept or between two words using the network above, though this correlation may or may not be the conditional probability.

In 2002, Google engineers constructed correlations of documents, concepts, and keywords using Bayesian networks, converting millions of keywords into clusters of concepts called Phil Clusters. Initially, this project did not consider any specific application background; it was motivated by the thought that concept extraction would be beneficial for future information processing. Phil Cluster's earliest application was extended matching of advertisements, which came during the several months following its completion. Because only the keyword to document relationship was taken into consideration in the early stages of Phil Cluster without much attention to contexts, this concept clustering was

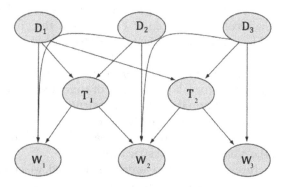

FIGURE 23.3: A Bayesian network describing documents (D), topics (T), and keywords (W).

applied too widely. For example, computers and stocks might be clustered, and cars might be grouped along with presidents' names. This severely impacted various Google project teams' acceptance of Phil Cluster. In 2004, Google started to restructure Phil Cluster with a data size more than a few hundred times the original. Keywords similarity calculation was also extended from using their co-occurrence in the same document to co-occurrence with context at the paragraph level, meaning the use of information at a finer level. Because it was a restructuring, the project was renamed RePhil. RePhil clustered about 12 million words into millions of concepts. Each concept usually contains dozens to hundreds of words. Since the quality of RePhil was greatly improved, many applications from ads to searching employed its results. Over time, few people know about Phil. The Bayesian network structure of RePhil is much more complex than that of Phil, but the underlying principles are similar.

23.3   Extended reading: Training a Bayesian network

Suggested background knowledge: probability theory.

In order to use a Bayesian network, we need to first specify its structure. For simple problems, such as the above example about cardiovascular disease, we can simply defer to the experts to provide its structure. However, for slightly more complicated problems, we cannot rely on structures produced manually. Here, we need to turn to machine learning.

An optimized Bayesian network needs to ensure that the ordering it generates (for example, "family history – high blood fats – cardiovascular disease" is an ordering) is the most likely. Measured by probability, this ordering should have the largest posterior probability. Of course, there are many paths that could generate an ordering, and global optimization theoretically requires exhaustive search that takes into account every path. However, this kind of computation's complexity is NP-hard. Therefore, we generally adopt a greedy algorithm, which entails finding the local optimal step at each step in the given direction. Of course, this would lead to local optimization and lead us astray from the global optimal solution. One way is to employ the Monte Carlo method to slightly randomize the gradient in iteration steps to prevent being trapped in local optimization. This method requires a large amount of computation. In recent years, the new method is using information theory to calculate the mutual information between two nodes, only keeping direct connections between nodes with large mutual information, and then conduct an exhaustive search on the simplified network to find the globally optimal structure.

After the structures of Bayesian networks are learned, we need to specify the weights of curves between nodes. Suppose these weights are measured in conditional probabilities. In order to accomplish this, we need some training data. We need to optimize the parameters of this Bayesian network to maximize the probability of observing these data (i.e., posterior probability) $P(D \mid \theta)$. This process is the aforementioned expectation-maximization (EM) process.

When computing the posterior probability, we are computing the joint probability $P(X, Y)$ of the condition X and the result Y. Our training data will provide some constraints between $P(X, Y)$, and the learned model needs to satisfy these constraints. Recalling our discussion in Chapter 20 Don't put all your eggs in one basket: The principle of maximum entropy, this model should be a maximum entropy model that satisfies all given conditions. Hence, we can employ the training methods relevant to maximum entropy models here.

Lastly, we need to point out that structure learning and parameter learning are often conducted alternately. In other words, parameter optimization usually comes first, then structure optimization, then parameter optimization again until convergence or a model with errors small enough is obtained.

Readers particularly interested in Bayesian networks can refer to the paper jointly published by Zweig and Bilmes:

http://ssli.ee.washington.edu/~bilmes/pgs/short_date/html

If you wish to learn more about this area systematically, I recommend a volume by Professor Daphne Koller from Stanford University, *Probabilistic Graphical Models: Principles and Techniques*. However, this book has over one thousand pages and is very expensive, so it is only suitable for professionals.

23.4   Summary

Mathematically, a Bayesian network is a weighted directed graph, and a generalization of a Markov chain. It has overcome Markov chains' restriction of mechanical linearity. Any related things can be clustered in a Bayesian network. Hence, Bayesian networks have many applications. Aside from text classification and concept extraction that we have already introduced, Bayesian networks are also widely applied in biostatistics, image processing, decision support systems, and game theory. Though a Bayesian network can be described in simple and easily comprehensible words, the model it produces can be highly complex.

Chapter 24

Conditional random fields, syntactic parsing, and more

Conditional random fields are an effective model for computing joint probability distributions, whereas syntactic parsing sounds like a lesson in English class. How are these two related? We will start from the evolution of syntactic parsing.

24.1 Syntactic parsing—the evolution of computer algorithms

Sentence parsing in natural languages generally refers to analyzing a sentence according to syntax and constructing this sentence's parse tree, i.e., syntactic parsing. Sometimes it can also refer to analysis of each sentence's component semantics, which results in a description of the sentence's semantics (such as a nested frame structure, or a semantic tree), i.e., semantic parsing. The topic of our discussion in this chapter is the first kind, which is the syntactic parsing of sentences. The research in this field used to be influenced by formal linguistics and used rule-based methods. The process of constructing this parse tree was to continually use the rules to merge the end nodes step-by-step, until the root node, which is a whole sentence. This is a bottom-up method. Of course, it could also proceed from the top down. Regardless of its direction, there is an inevitable problem, which is that it is impossible to choose the correct rules all at once. One misstep requires retracing of many steps. Therefore, both of these methods' computations are extremely complex, and it is impossible to analyze complicated sentences.

After the 1980s, computational linguist, Eugene Charniak, professor of computer science and cognitive science at Brown University, computed the probabilities of grammar rules. When choosing grammar rules, he stuck with one principle - maximize the probability of the sentence's parse tree. This seemingly simple and direct approach significantly reduced the amount of computation required for syntactic analysis while it greatly increased its accuracy. Charniak built a bridge between mathematics and syntactic analysis. The second person to build this bridge was Mitch Marcus's student, Adwait Ratnaparkhi.

Eric Brill, Ratnaparkhi's ex-group mate in Mitch Marcus's lab and former VP of research at eBay, highly appraises the talent of his junior group mate. "Ratnaparkhi is extremely intelligent. From my interactions with him, he is extremely good at finding practical ways to implement theoretical ideas efficiently." Ratnaparkhi proposed a new angle from which to view syntactic analysis problems - he saw syntactic analysis as a process of bracketing.

We will demonstrate Ratnaparkhi's method using this quote from Chapter 3 as our example:

> The Fed Chairman Ben Bernanke told media yesterday that $700 billion bailout funds would be lent to hundreds of banks, insurance companies and auto-makers.

We first segment consecutive characters in the sentence into words as shown below:

> The Fed | Chairman | Ben Bernanke | told | media | yesterday | that | $700 billion | bailout | funds | would | be lent to | hundreds | of | banks |, | insurance companies | and | auto-makers |

Then, we scan these words from left to right, and bracket words to form phrases:

> (The Fed Chairman) | Ben Bernanke | told | media | yesterday | that | $700 billion | (bailout funds) (would be lent to) (hundreds of) (banks, insurance companies and auto-makers)

Next, we assign the components syntactic non terminal labels, e.g., NP(noun phrase) to "The Fed Chairman", and then repeat the above bracketing process using sentence components as the unit, as shown below:

> [(The Fed Chairman) Ben Bernanke] told | media | yesterday | [that $700 billion | (bailout funds)] (would be lent to) [(hundreds of) (banks, insurance companies and auto-makers)]

This process is done recursively until the whole sentence is dominated in a large bracket. Each bracket is a sentence component, and the nested relationship between brackets is the relationship between sentence components at different levels.

Each time Ratnaparkhi scans every word (or sentence component) from left to right, he only needs to take one of three actions listed below:

A1. Whether to start a new left bracket. For example, "The Fed" is the start of a new bracket.

A2. Whether to stay in the current bracket. For example, at "insurance companies," we stay in the bracket.

A3. Whether to end a bracket, i.e., close the right bracket. For example, "funds" is the end of a bracket.

In order to determine which action to take, Ratnaparkhi constructed a statistical model $P(A \mid \text{prefix})$. Here, "A" represents the action and "prefix" refers to all of the words and sentence components from the beginning of the sentence to now. Lastly, Ratnaparkhi implemented this model using a maximum entropy model. Of course, he also used a statistical model to predict the types of sentence components.

This method is extremely fast. After each scanning, the number of sentence components decreases proportionally. Hence, the number of scans is the logarithmic function of the sentence's length. It is easy to prove that the entire algorithm is proportional to the length of the sentence. In terms of algorithm complexity, the computation time for this algorithm is already the most optimal.

Ratnaparkhi's method appears very simple, but it is an amazing feat to be able to come up with it. (Hence, we see that good methods are often simple in form.) It would not be an overstatement to say that he was a key figure in making the connection between sentence syntactic analysis and mathematical modeling. From then on, most syntactic analysis approaches turned from heuristic searching to bracketing. The accuracy of syntactic analysis largely depends on the quality of the statistical model and the effectiveness of feature extraction from the prefix. Michael Collins, the Pandit Professor of Computer Science at Columbia University today, once wrote one of the best dissertations in natural language processing when he was a PhD candidate under the supervision of Prof. Mitch Marcus, because Collins conducted great research in feature extraction.

From the emergence of statistical linguistics and statistical language models in the 1970s, to the integration of mathematics and syntactic analysis through statistical models in the 1990s, syntactic analysis of "standard" sentences, such as sentences from *The Wall Street Journal*, has an accuracy rate of over 80%. This has basically reached the level of an average person. After the turn of the century, with the spread of the Internet, content created by netizens replaced rigorous writings by professionals as the main constituent of writings that reach most readers. For those "less-than-well-written" sentences or even sentences with major grammatical errors, the accuracy rate of a conventional syntactic parser - even the one developed by Collins - struggled to reach 50%.

Luckily, in many natural language processing applications, we do not need to analyze sentences too thoroughly. Only shallow parsing would suffice in most cases. For example, we only need to find the main phrases in a sentence and the relationships between them. Hence, scientists shifted the focus of syntactic analysis to shallow parsing. After the 1990s, with the increase of computers' computational power, scientists adopted a new mathematical modeling tool - conditional random fields, which significantly increased the accuracy rate of shallow parsing to as high as 95%. This expanded the applications of syntactic analysis to many products such as machine translation.

24.2 Conditional random fields

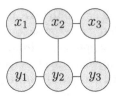

FIGURE 24.1: In a hidden Markov model, output is only dependent on the state.

In a hidden Markov model, $x_1, x_2, \ldots, x_n$ represent the observation sequence, and $y_1, y_2, \ldots, y_n$ represent the hidden state sequence. Here, x_i only depends on the state y_i that generates it, i.e., y_{i-1} or y_{i+1}. Obviously, in many applications, the observed value x_i may be related to the states before and after. Taking into account x_i and y_{i-1}, y_i, y_{i+1}, the corresponding model is as follows:

This model is a conditional random field.

The conditional random field is an expansion of the hidden Markov model. It retains certain features of the hidden Markov model. For example, the state sequence $y_1, y_2, \ldots$ is still a Markov chain. More broadly speaking, a conditional random field is a special probabilistic graph model. In this graph, each node represents a random variable, such as x_1 and y_1. The link between nodes represents their mutually dependent relationship, which is generally described with a probability distribution such as $P(x_1, y_1)$. Its uniqueness lies in that the variables conform to the Markov hypothesis. In other words, each state's transfer probability only depends on its adjacent states. This is similar to another type of probability graph model we introduced previously, Bayesian networks. The difference between the two is that while conditional random fields are undirected graphs, Bayesian networks are directed graphs.

In most applications, the nodes of a conditional random field can be divided into two disjoint sets, X and Y, representing the observations and output variables, respectively (see Figure 24.2). The conditional random field's quantified model is these two sets' joint probability distribution model $P(X, Y)$

$$P(X, Y) = P(x_1, x_2, \ldots, x_n, y_1, y_2, \ldots, y_m) \tag{24.1}$$

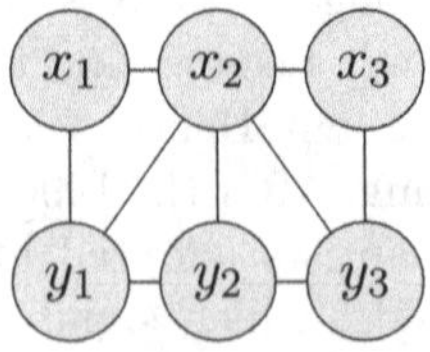

FIGURE 24.2: A typical conditional random field.

Because there are too many variables in this model to obtain enough data to estimate directly with the law of large numbers, we can only use its marginal distributions, such as $P(x_1)$, $P(y_2)$, $P(x_1, y_3)$, to find a probability distribution function that satisfies all of the constraints. Of course, there may exist more than one such function (it is generally so). According to the maximum entropy principle, we hope to find a maximum entropy model that satisfies all the marginal distributions. We have mentioned before that this model is an exponential function. Each marginal distribution corresponds to a feature in the exponential function f_i. For example, the marginal distribution feature of x_1 is as follows:

$$f_i(x_1, x_2, \ldots, x_n, y_1, y_2, \ldots, y_m) = f_i(x_1) \tag{24.2}$$

This feature demonstrates that it only depends on x_1 but no other variables. If the values of some variables corresponding to a certain feature function are zero, then this feature function has no effect on these variables. Applying these features in the model, we obtain the following,

$$P(x_1, x_2, \ldots, x_n, y_1, y_2, \ldots, y_m) = \frac{e^{f_1 + f_1 + \cdots + f_k}}{Z}, \tag{24.3}$$

where k is the total number of features and Z is the normalization constant that makes P a proper probability. Using shallow parsing as an example, we can demonstrate how a conditional random field model is trained and how it works.

Suppose X represents what we see. In shallow parsing, it is the words in a sentence, their parts of speech, and so forth. Y represents what we need to derive, which is the grammatical composition, including noun phrases, verb phrases, time phrases, and so on. Take the simple example "Romeo loves Juliet" that we have used before, we can illustrate the principle of this shallow parsing model. Here, the observed sequence is Romeo/noun, loves/verb, Juliet/noun, and the state sequence we hope to find is "noun phrase, verb phrase." We will use the channel model to demonstrate this analysis process, as shown in Figure 24.3:

Y=noun phrase, verb phrase. X=Romeo/noun, loves/verb, Juliet/noun.

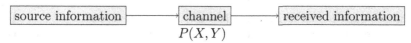

FIGURE 24.3: A channel model of sentence syntactic parsing.

In the parse tree in Figure 24.4, its features are the different nodes, ordered combinations of same-level nodes, combinations of different-level nodes, and so on. Here, ordered combinations of same-level nodes could be "Romeo - loves," "verb - noun," etc. Combinations of different-level nodes are actually subtrees. For example, "noun phrase" can be rewritten as "noun," "verb phrase" can be

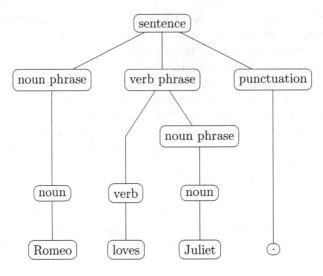

FIGURE 24.4: Syntactic parse tree for "Romeo loves Juliet."

rewritten as "verb, noun phrase," etc. After considering some features, we can use the following equation to compute this conditional random field's probability:

$$P(\text{Romeo/noun, loves/verb, Juliet/noun, noun phrase, verb phrase})$$
$$= \exp\{f_1(\text{Romeo, noun})$$
$$+ f_2(\text{loves, verb}) + f_3(\text{Juliet, noun})$$
$$+ f_4(\text{Romeo, noun, noun phrase}) + f_5(\text{noun, noun phrase})$$
$$+ f_6(\text{loves, verb, Juliet, noun, verb phrase})$$
$$+ f_7(\text{noun phrase, verb phrase}) + f_8(\text{noun phrase})$$
$$+ f_9(\text{verb phrase}) + f_{10}(\text{verb, noun, verb phrase})\}$$

$$(24.4)$$

Here, the parameters of feature function $f()$ can be obtained using the maximum entropy model training method introduced previously.

The earliest researchers to adopt conditional random fields in shallow parsing of sentences were Fei Sha, then a PhD student at the University of Pennsylvania, and one of his mentors, Fernando Pereira. They continued Ratnaparkhi's legacy of only parsing the first level of the sentence, which is the natural grouping of words into phrases. Because of the improved statistical model, the accuracy rate of shallow parsing is as high as 94%. In natural language processing, this is a very impressive accuracy rate.

In 2005, Dr. Kai-Fu Lee joined Google and took over the Chinese Web Search Project from Huican Zhu and me. I then led a team to develop a general syntactic parser to serve the whole company, GParser (G stands for Google). Our approach was similar to that of Sha and Pereira (Pereira had not joined Google then). What was different was that we did not stop at the first level of analysis, but parsed all the way to the top of the sentence parse tree like Ratnaparkhi did. The basic model was still a conditional random field.

For each level of analysis, we constructed a model $P(X, Y)$. Here, X is the words in a sentence $w_1, w_2, \ldots, w_n$, parts of speech $pos_1, pos_2, \ldots, pos_n$, names of each level's grammatical components $h_1, h_3, \ldots, h_m$, and so on. Y is the action (left bracket, stay in the bracket, and right bracket) and names of the new level's grammatical components. The expanded equation is as follows:

$$P(w_1, w_2, \ldots, w_n, pos_1, pos_2, \ldots, pos_n, h_1, h_2, \ldots, h_m, Y) \qquad (24.5)$$

This seemingly all-inclusive model actually has a big engineering problem. First, this model is highly complex and very difficult to train. Second, the number of various constraint combinations is astronomical. Even with Google's data, none of the combinations would appear more than a few times. Hence, we need to solve the two following problems.

The first step is approximation. We need to break down various constraint combinations $w_1, w_2, \ldots, w_n, pos_1, pos_2, \ldots, pos_n, h_1, h_2, \ldots, h_m$ into many subsets. For example, a subset of the last two words w_{n-1}, w_n or of the last two sentence components h_{m-1}, h_m. Reliable statistical relationships can be found between each subset and the action to be predicted (and the names of higher level grammatical components).

Second, in the training data, we need to set every statistical correlation with sufficient statistics as a constraint. Our goal is to find a maximum entropy model that satisfies all constraints.

Thus, we can use the maximum entropy model method to obtain this syntactic parser.

For a syntactic parser like this, its accuracy rate for analyzing sentences in a webpage has already reached that of an analysis of *The Wall Street Journal* published in Fei Sha's dissertation. It is worth mentioning that compared to similar tools in academia, this parser can process very "dirty" webpage data. Therefore, it was later applied in many Google products.

24.3   Conditional random field applications in other fields

Of course, the application of conditional random fields is not limited to natural language processing. Even in many traditional industries, conditional random fields also prove to be surprisingly effective.

In urban areas in the US, crime is a big problem for both law enforcement and for the residents. The best solution is not solving the case after a crime occurs, but preventing criminal behavior from happening. Of course, this is easier said than done. The cities are very big, and population densities very high. Many unexpected events can occur at any time, making it difficult to predict where the next crime would take place. In the past, the police patrolled the streets and could only prevent a crime from occurring when the criminal happened to be nearby. However, such "coincidence" was rare. Now, law enforcement can analyze big data with mathematical models and effectively predict where and when a crime may occur in a city, then carry out targeted patrolling to stop crimes.

Los Angeles Police Department (LAPD) was the first to come up with this idea, and entrusted this task to the University of California, Los Angeles (UCLA). UCLA constructed a conditional random field-based mathematical model, using more than 13 million criminal cases that happened in the past more than 80 years. This model predicts the probabilities of the times, locations, and categories of crimes that may occur. If we use l, t, and c to represent location, time, and category of crimes, respectively, then this mathematical model estimates the probability $P(l, t, c)$.

The time t in the model above is calculated in minutes. In order to accurately describe the location information, l, LAPD divided thousands of square kilometers of the city into 5 meter × 5 meter grids. As for c, LAPD has its own categorization method. Of course, this model needs to take many other factors into account, such as traffic flow, events (athletic games, concerts, etc.), the weather, unemployment rate, and so forth. We can represent these factors with a vector $X = x_1, x_2, \ldots, x_n$. Now, with these factors added into our consideration, the probability to be predicted is in fact the distribution of location, time, and crime category given the condition of factors X: $P(l, t, c|X)$. Using the conditional probability equation to expand it, we obtain the following:

$$P(l, t, c|X) = \frac{P(l, t, c, X)}{P(X)} \tag{24.6}$$

Note that in this equation, the denominator $P(X)$ on the right-hand side is estimated from historical data, which can be considered as a known variable. Hence, this model's key component is the numerator on the right-hand side, $P(l, t, c, X)$. Compare this expression to that on the left side of equation (24.1), we can see that the two expressions are very similar in form. Therefore, we can use a conditional random field to estimate this probability.

The mathematical model that UCLA professors adopted was a conditional random field. First, they used a priori probability distribution as the starting point of the whole model. They called this priori probability distribution the background distribution. In other words, this is the default distribution of criminal rate without having any knowledge related to time and place. Of course, the background distribution is not too accurate. Hence, UCLA professors extracted various features according to historical data, such as the relationships between shows and theft, between robbery and the flow of people, between athletic games and drunk driving, etc. We can obtain a model similar to (24.4) by combining these features, and then predict potential crimes using this model. Obviously, training this model needs a large amount of data, so LAPD provided all 13 million cases on record to train the model well. This model was later proven to be able to predict potential crimes in the future to a certain degree, and successfully reduced the crime rate by 13%. This invention was hailed as one of the best inventions of the year by *Time* magazine in 2011.

24.4   Summary

Conditional random fields are a highly flexible statistical model that makes effective predictions. This chapter emphasizes its application in natural language processing, especially in sentence parsing, but its application goes far beyond this field. Conditional random fields have highly successful applications in areas including pattern recognition, machine learning, biostatistics, and even crime prevention.

Similar to maximum entropy models, conditional random fields are simple in form but complex in implementation. Fortunately, there are enough open source softwares available today to satisfy the needs of most engineers.

Chapter 25

Andrew Viterbi and the Viterbi algorithm

Speaking of Andrew Viterbi, perhaps few people outside the communication industry know about him. Nonetheless, most people in computer science and telecommunications know about the Viterbi algorithm named after him. The Viterbi algorithm is the most common algorithm in modern digital communications. It is also the decoding algorithm widely used in natural language processing. It is not an exaggeration to say that Andrew Viterbi is one of the scientists whose work has had the most influence on our lives today. Andrew Viterbi and Irwin Jacobs co-founded Qualcomm, a leading company of semiconductor and telecommunications in the world that dominated CDMA-based 3G mobile communication standards. Today, Qualcomm is still a leader in mobile communication in the 4G and 5G eras.

25.1   The Viterbi algorithm

The first time I heard about Viterbi's name was in 1986, when I was studying the Viterbi algorithm in graph theory. That year, he and Dr. Jacobs had just founded their second company, Qualcomm (registered in 1985). Back then, Viterbi was only known in academia, specifically in the fields of communication and computer algorithm. The first time I used the Viterbi algorithm was in 1991, when I was conducting research in speech recognition. That year, Qualcomm had proposed and improved the foundation of today's 3G communication - CDMA technology. When I first met Dr. Viterbi at Johns Hopkins University in the summer of 1997, he was already a world-renowned communication industry giant. That year, he attended the annual meeting as the advisor to the Center for Language and Speech Processing (CLSP), and heard us report the work at CLSP. What he cared about the most was the business prospects of speech recognition. Today, he can see them.

Dr. Viterbi is an Italian-American Jewish immigrant. His original name was Andrea Viterbi. Since Andrea sounds like a woman's name in English, he changed his name to Andrew. From his graduation from MIT to when he was 33 years old, Viterbi's professional life was limited to academia. He served as an engineer at the major U.S. defense contractor and industrial corporation Raytheon, the famous Jet Propulsion Lab (JPL), and then completed his

PhD at the University of Southern California. He then taught at UCLA and UCSD, in the then-emerging discipline of digital communications. Several years later, in 1967, he invented the Viterbi algorithm.

Returning to our main topic, let us take a look at the famous algorithm named after Viterbi. The Viterbi algorithm is a special but widely applied dynamic planning algorithm (see previous chapters). Dynamic planning algorithms can be used to solve the shortest path problem in any graph. The Viterbi algorithm was proposed for a specific type of directed graph, the lattice graph. It is so important because any problem that can be described using a hidden Markov model can use the Viterbi algorithm to decode. Hence, it is applied in fields including digital communications, speech recognition, machine translation, and pinyin-to-hanzi (Chinese characters) conversion. Now, we will use a familiar example, pinyin-to-hanzi conversion in Chinese input methods, to show the Viterbi algorithm.

Suppose the user inputs the following pinyins (phonetics) $y_1, y_2, \ldots, y_N$, which correspond to Chinese characters $x_1, x_2, \ldots, x_N$ (although actual input method editors use words as the unit of input, we will use characters as the unit to explain the Viterbi algorithm for the sake of simplicity and clarity). According to tools introduced previously:

$$
\begin{aligned}
x_1, x_2, \ldots, x_N &= \arg \max_{w \in W} P(x_1, x_2, \ldots, x_N | y_1, y_2, \ldots, y_N) \\
&= \arg \max_{w \in W} \prod_{i=1}^{N} P(y_i | x_i) \cdot P(x_i | x_{i-1})
\end{aligned}
\tag{25.1}
$$

The input (visible) sequence is $y_1, y_2, \ldots, y_N$, and the hidden sequence that generates it is $x_1, x_2, \ldots, x_N$. We can describe such a process using the graph below (Figure 25.1):

This is a relatively simple hidden Markov chain, without any state jumps or state loops. $P(x_1 | x_{i-1})$ is the transfer probability between states, and $P(y_i | x_i)$ is each state's generating probability. Now, the output of each state in this Markov chain is fixed, but the values of these states can vary. For example, a character with the output pronunciation "zhong" could be "中" (middle) or "种" (plant), or another character. We can think more abstractly and use the symbol x_{xj} to represent the jth possible value of state x_i. Expanding each state according to its different values, we obtain the following lattice graph in Figure 25.2:

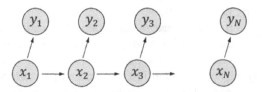

FIGURE 25.1: The hidden Markov model suitable for the Viterbi algorithm.

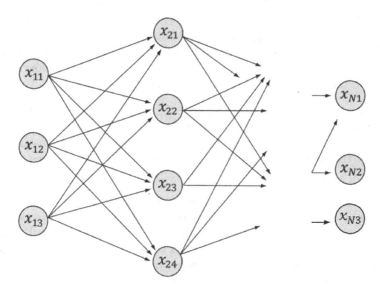

FIGURE 25.2: The lattice graph.

In the above graph, each state has 3 to 4 values, but of course they can have any number of values in real applications.

Any path from the first state to the last one has the potential to generate the output sequence of observed events Y. Of course, the probabilities of these paths are different, and our task is to find the most likely path. For any given path, we can use Equation (25.1) to calculate its probability, and this is not too difficult. However, there are a large number of such path combinations, which would make the number of sequences increase exponentially. Each tone-less pinyin in Mandarin Chinese corresponds to about 13 national standard characters on average in the GB character set. Suppose the sentence length is 10 characters, then the number of combinations is $13^{10} \sim 5 \times 10^{14}$. Suppose computing the probability of a path requires 20 multiplication operations (or additions, if the programmer is very intelligent), it would be 10^{16} computations. Today's computer processors* can compute 10^{11} times per second, which means that the computations would take about a day. However, in telecommunication or speech recognition, there are thousands of thousands of states for every sentence, so the exhaustive method above is obviously unsuitable. Hence, we need an algorithm that is proportional to the number of states. This algorithm was first proposed by Dr. Viterbi in 1967, and was named the Viterbi algorithm.

The fundamentals of the Viterbi algorithm can be summarized as the following three points:

---

*Intel Core i7 Extreme.

1. If the most probable path (or the shortest path) P passes a point, such as x_{22} in the graph, then the path Q from the starting point S to x_{22} on this path P must be the shortest path between S and x_{22}. Otherwise, replace Q with R, the shortest path from S to x_{22}, would make a new path that is shorter than P. This would be self-contradictory.

2. The path from S to ending point E must pass through a state at time i. Suppose there are k states at time i. If we record the shortest paths from S to all k nodes of the ith state, then the ultimate shortest path must pass through one of them. Thus, at any given time, we only need to consider a limited number of shortest paths.

3. To sum up the two points above, if the shortest path of every node from S to state i has already been found when we go to state $i+1$ from state i, and these paths are recorded on the nodes, then when computing the shortest path from S to a node x_{i+1} of state $i+1$, we only need to consider the shortest paths of all k nodes from S to the previous state i and the distances from these k nodes to x_{i+1}, j.

Based on these three points, Viterbi came up with the following algorithm:

First, starting from point S, for the first state x_1's nodes, we can suppose that there are n_1 of them and compute the distances from S to them, $d(S, x_{1i})$. Here, x_{1i} represents any node of state 1. Because there is only one step, all of these distances are the shortest paths from S to these nodes, respectively (see Figure 25.3).

The second step is the key to understanding the entire algorithm. For the second state x_2's nodes, compute the shortest distances from S to them. We know that for a specific node x_{2i}, the path from S to it can pass through any

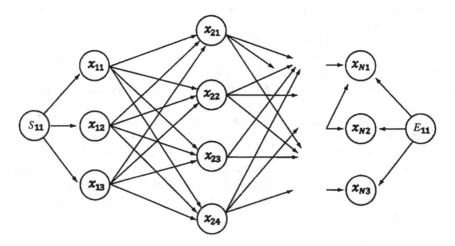

FIGURE 25.3: A state that the path from the starting point to the ending point must pass at time i.

node x_{1i} of state 1's, e.g., n_1. Of course, the length of the corresponding path is $d(S, x_{2i}) = d(S, x_{1j}) + d(x_{1j}, x_{2i})$. Since j has n_1 possibilities, we need to compute them one-by-one and find the minimum:

$$d(S, x_{2i}) = min_{i=1}^{n_1} d(S, x_{1j}) + d(x_{1j}, x_{2i}) \tag{25.2}$$

Computing each node of the second state requires n_1 multiplication operations. Suppose this state has n_2 nodes, computing the distances from S to all of them needs $O(n_1 \cdot n_2)$ calculations.

Next, we go from the second state to the third state following the approach above, until we reach the last state. Now, we have obtained the shortest path of the entire lattice graph, from the beginning to the end. In each step, computational complexity is proportional to the product of n_i and n_{i+1}, the numbers of nodes of two adjacent states S_i and S_{i+1}, i.e., $O(n_i \cdot n_{i+1})$. Suppose the state with the largest number of nodes in this hidden Markov chain has D modes, which is to say that the lattice graph's width is D, then the complexity of any step does not exceed $O(D^2)$. Since the lattice graph's length is N, the Viterbi algorithm's complexity is $O(N \cdot D^2)$.

Returning to the input method editor problem discussed above, the amount of computation is about $13 \times 13 \times 10 = 1690 \approx 10^3$. This is qualitatively different from the original 10^{16}. More importantly, the Viterbi algorithm is proportional to the length. Regardless of telecommunication, speech recognition, or typing, data is inputted in a stream, so as long as the time it takes to process each state is faster than the speed of speaking or typing (this is easily achieved), the decoding process is always in real time no matter how long the input is. This is where we see the beauty of mathematics!

Although the fundamental algorithm of digital communication and natural language processing has a simple principle that any communication or computer science student can learn in two hours, it was a great achievement to be able to come up with this fast algorithm in the 1960s. Dr. Viterbi established his reputation in digital communication with this algorithm. However, he did not stop at inventing the algorithm; he worked very hard to promote and popularize it. To this end, Viterbi did two things: first, he gave up the algorithm's patent; second, he founded Linkabit with Dr. Jacobs in 1968 and sold the algorithm in a chip to other communication companies.

Until now, Dr. Viterbi had already gone much farther than most scientists up until that point. However, this was only the first step of his brilliant life. In the 1980s, he devoted himself to applying CDMA technology to wireless communication.

25.2   CDMA technology: The foundation of 3G mobile communication

Since Apple started selling iPhones, 3G phones and mobile internet became hot topics in technology and business. The crucial communication technology here is code-division multiple access (CDMA) technology. There are two people who contributed the most to the invention and popularization of CDMA technology - Austrian-American, Hedy Lamarr, and Viterbi.

Lamarr was probably the most beautiful scientist in history. In fact, her main occupation was acting, and she invented the frequency modulation communication technology in her spare time. Lamarr had been learning dance and piano since she was 10 years old, and entered the entertainment industry later on. When playing the piano, Lamarr thought of encoding the communication signal with different frequencies of piano keys. Knowing the frequency modulation sequence, the receiver can decode the signal. However, the signal cannot be decoded without knowing the frequency. For example, if you know Chopin's *Heroic Polonaise (Op. 53)*, then you would recognize it when it is being played. Otherwise, it would be a series of random notes. Lamarr and her neighbor, the composer George Antheil, invented a frequency modulation communication technology called the "secure communication technology." Here, the communication signal's carrier frequency hops quickly; the sender and the receiver only need to agree on a sequence (usually a pseudo-random number sequence) beforehand. The eavesdropper cannot do anything with the information without knowing this sequence. Lamarr first used the frequencies of 88 piano keys as the carrier frequency, and put the agreed-upon frequency modulation sequence on the piano roll, then the carrier frequency would hop according to perforations' positions on the piano roll.*

This frequency modulation technology is the predecessor of CDMA. It was granted a patent in 1941. The US Navy originally wanted to use this technology to implement radio-controlled torpedoes that could not be detected by the enemy, but the plan was aborted because of some objections. Soon, WWII ended. This technology was not implemented until 1962. During the Vietnam War, the Vietnamese military discovered that US pilots who were shot down were able to request help using a device with an undetectable frequency. After seizing the device, they were unable to understand its principle or decode its signal, so they gave it to the Chinese advisory group aiding Vietnam at the time. The advisory group had some communication experts, including my mentor at Tsinghua University, Professor Wang Zuoying. The Chinese experts found that this kind of device transmits coded signals on a very wide band with a very low power. For those trying to intercept the signals, the power is too low to be captured. Even when the signal is captured, the intercepter would not be able to decode it. However, the receiver can obtain the information by accumulating the low powers and decode it with the key.

---

*A piano roll is a continuous paper roll that automatically operates a player piano, similar to the paper roll used in the earliest computers. Perforations on a piano roll represent different nodes, so the track bar can read the notes and play the piano accordingly. This is similar to today's MIDI files.

This kind of transmission is conducted on a relatively wide extended band, so it is called spread-spectrum transmission. Compared to fixed frequency transmission, it has three advantages:

1. The anti-interference ability of spread-spectrum transmission is very strong. In the 1980s, listening to foreign radios was prohibited in China. However, this was very hard to enforce because wireless signals filled the air and were easy to tune into. Hence, what the Chinese government could do was create noise interference to those fixed radio frequencies. For spread spectrum transmission, this is almost impossible because the government cannot interfere in all of the frequency bands.

2. Spread-spectrum transmission signals are very difficult to intercept. We have already discussed this.

3. Spread-spectrum transmission uses the bandwidth more efficiently. This is too complicated to explain in detail. Simply speaking, because of mutual interference of adjacent frequencies in fixed frequency communication, frequencies on the carrier's band have to be distributed sparsely, and the frequency bandwidth between two frequencies is wasted. Spread-spectrum communication does not waste as much frequency band because of its strong anti-interference ability.

Although spread spectrum technology and frequency modulation technology were used in the military as early as the 1960s, they did not enter the non-military market until the 1980s. The need for mobile communication was the main reason for this. In the 1980s, mobile communication began to develop rapidly. Soon, people realized that there were not enough frequency bands, and new communication technologies were needed. Furthermore, as for CDMA technology, Viterbi's contribution was indispensable.

Before CDMA, mobile communication had used two kinds of technologies: frequency-division multiple access (FDMA) and time-division multiple access (TDMA).

As the name suggests, FDMA divides the frequency so that each line of communication has its own frequency. This is the principle of walkie-talkies. Because adjacent frequencies would mutually interfere, each channel needs to have enough bandwidth. If the number of users increases, the total bandwidth has to increase as well. We know that the bandwidth resource in the air is limited. That means that a balance between the number of users and the transmission rate must be reached.

TDMA divides the same frequency band by time. Each user takes only up $1/N$ of the transmission time of a fixed sub-frequency band. Thus, many people can use the same frequency band simultaneously. The second generation of mobile communication standard was TDMA-based.

We have mentioned before that spread-spectrum transmission uses the frequency band more efficiently than fixed frequency transmission. Therefore, if we

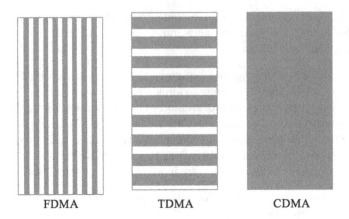

FDMA   TDMA   CDMA

FIGURE 25.4: FDMA, TDMA, and CDMA's usage of frequency bands and time; the shaded area is the utilized part, the rest is the unused part.

use many sub-divided frequency bands to transmit information simultaneously, we should be able to utilize the bandwidth more efficiently. This way, we can increase the number of users, or increase the speed of transmission for everyone given a set number of users. This way, we can increase the number of users in the network and increase the transmission rate (Figure 25.4).

The two main wireless carriers in the US are AT&T and Verizon. While the former's base station density is not lower than that of the latter, nor is the former's signal weaker than that of the latter, the former's call quality and transmission speed lag behind those of the latter. This is because AT&T's main network relies on the old TDMA technology, but Verizon's network is completely CDMA-based.

Some readers may ask: if one transmitter takes up many frequency bands, then wouldn't multiple transmitters that send information simultaneously be in conflict with one another? This is not a problem because each sender has his or her own encrypted code, and the receiver will filter out signals that cannot be decoded, leaving only the signal that corresponds to the shared code. Because this method uses different codes to allow multiple simultaneous transmissions, it is called code-division multiple access.

The first company to apply CDMA technologies to mobile communication was Qualcomm. From 1985 to 1995, Qualcomm developed and perfected the CDMA communication standard CDMA1, and established the world's first industry-leading 3G communication standard CDMA 2000. Later, together with communication companies from Europe and Japan, Qualcomm established the world's second 3G standard WCDMA. In 2007, as a mathematician and a computer scientist, Dr. Viterbi was awarded the highest honor in science and engineering - the National Medal of Science.*

---

*http://www.nsf.gov/od/nms/recip_details.cfm?recip_id=5300000000446

Perhaps because of Viterbi's strong technical background, Qualcomm is a purely technology-driven company. Although it is the world's largest 3G cell-phone processor company and the semiconductor company with the largest market value today, Qualcomm does not have semiconductor manufacturing - only its research, development, and design. A large portion of Qualcomm revenue comes from license fees of intellectual properties. Sometimes, if a company's technologies are too advanced but the market is not ready, they may not help the company win in business and this was the case of Qualcomm in 2G mobile phone era. Even though its technologies were better than most of its European competitors, the advantage was still lost in the market at that time because fast data transmission was not a priority for 2G users. However, Qualcomm's technical advantage ensured its ruling position in the third generation of mobile communication. It not only is a leader of mobile communication in the 4G era, but also once beat Intel as the world's largest semiconductor company in the largest market cap.

If we count Viterbi among the mathematicians, then he may be the second richest mathematician in history (the richest is no doubt the founder of Renaissance Technologies, James Simons). Viterbi is one of the University of Southern California's (USC) biggest donors.* USC's engineering school was named after him. Viterbi's wealth came from his success in turning technology into business.

25.3   Summary

Most scientists in the world would be satisfied if the results of their research are recognized by peers in the academy, and they would be overjoyed if their research can be widely used in industry and generate profit. However, very few scientists can apply their research accomplishments to real products by themselves, because entrepreneurship is difficult for scientists. There are some exceptions, like John Hennessy, who invented reduced instruction set computing (RISC) technologies and also successfully founded the company of the same name, and John Cioff, the father of asymmetric digital subscriber line (ADSL). Although these people had already accomplished a lot, they still only focused on the areas that they were good at rather than carrying out the revolution from the beginning to the end. However, what Viterbi did went far beyond this point. He not only brought the key inventions, but also applied them to make our world better. He even developed all of the supporting technologies and built entry barriers in industry. Many companies that tried to make a detour around Qualcomm's intellectual properties found it impossible, because Qualcomm has already patented everything.

---

*He donated $52 million to USC.

Chapter 26

God's algorithm: The expectation-maximization algorithm

We have discussed automatic document classification many times in previous chapters. This is because today's internet products and applications all need to use this technology. On the other hand, this technology can be applied to almost any kind of classification, such as user classification, term classification, product classification, even biological characteristics and gene classification. In this chapter, we will further introduce some automatic document classification technologies and use them to illustrate an important method in machine learning - the expectation-maximization (EM) algorithm. I call this algorithm "God's algorithm."

26.1   Self-converged document classification

We have already introduced two document classification algorithms in previous chapters. One of them uses pre-defined categories to classify new articles, while the other one has not such human-defined categories and clusters articles by comparing them in pairs from the bottom up. It first randomly assigns the centroids of document clusters, and then optimizes the clustering results by many iterations until the new centroids approximate to the real ones as close as possible (or converging to real clusters). Of course, this method still needs to use TF-IDF and the cosine distance between vectors. These have been discussed in previous chapters as well.

Self-converged classification can be simply explained as follows. Suppose there are N articles, corresponding to N vectors $V_1, V_2, \ldots, V_N$. We need to classify them into K categories, and their centroids are $c_1, c_2, \ldots, c_K$. We can consider both the vectors and the centroids as points in space. Of course, K could be a fixed number, such as 100 different article topics, or categories; it could also be a number that varies - for example, if we do not know the number of topics or categories beforehand, then we can classify articles into however many categories there are in the end. The classification steps are as follows:

1. Randomly select K points as the initial centroids $c_1(0), c_2(0), \ldots, c_K(0)$. In the following graph, the points belong to three different categories. We use black crosses to represent randomly appointed cluster centroids. See Figure 26.1.

FIGURE 26.1: The initial randomly generated three cluster centroids.

2. Compute the distances from each point to the cluster centroids, and classify the points into the closest cluster, as shown in the graph below (Figure 26.2).

3. Recompute the centroid of each cluster. Suppose point v is in one of the clusters, each point has multiple dimensions:

$$v_1 = v_{11}, v_{12}, \ldots, v_{1d}$$
$$v_2 = v_{21}, v_{22}, \ldots, v_{2d}$$
$$\ldots$$
$$v_M = v_{m1}, v_{m2}, \ldots, v_{md}$$

The easiest way is to use these cluster centroids $w = w_1, w_2, \ldots, w_m$ as the center of v. Its ith dimension value can be computed as follows:

$$w_i = \frac{v_{1i} + v_{2i} + \quad + v_{mi}}{m} \tag{26.1}$$

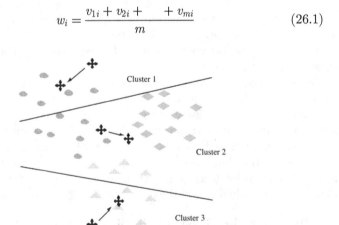

FIGURE 26.2: Recluster each point according to the centroids, and compute new centroids.

The new cluster centroids have different locations than the original ones. Figure 26.2 represents the location shifts from the old centroids to the new ones.

4. Repeat the above steps until the new centroids' displacements are very very small, i.e., convergence.

The clustering process will converge after a few iterations (see Figures 26.3 and 26.4). Of course, convergence of this method requires strict mathematical proofs, which will be shown later. This method does not require any artificial interference or prior experience; it is pure mathematical computations and will eventually accomplish document classification automatically. This is almost too good to be true. The reader might ask: can this really guarantee that the closest points are clustered together? If so, why?

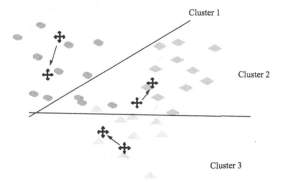

FIGURE 26.3: The results after two iterations.

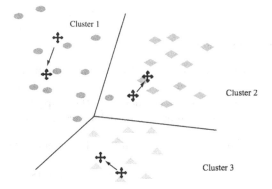

FIGURE 26.4: The results after three iterations, close to convergence.

26.2   Extended reading: Convergence of expectation-maximization algorithms

Suggested background knowledge: machine learning, pattern classification.

First, we need to ensure that the similarity measurement or the distance function is good enough so that it can make sure that points in the same categories are relatively close, while those in different categories are relatively far away from each other. Furthermore, an ideal clustering result should be like this: for all points in the same cluster, their average distance d to their centroid is short while the average distance to other cluster centroids D is long. We expect that after each iteration, the value of d decreases and the value of D increases.

If there are $n_1, n_2, \ldots, n_k$ points from the first cluster to the kth, in each cluster, the average distance from each point to the centroid is $d_1, d_2, \ldots, d_k$. Hence, $d = (n_1 \quad d_1 + n_2 \quad d_2 + \quad + n_k \quad d_k)/k$. Now, if the distance between the ith and the jth cluster centroids is D_{ij}, then $D = \sum_{i,j} \dfrac{D_{ij}}{k(k-1)}$. Taking cluster sizes (the number of points) into consideration, then D's weighted average equation should be

$$D = \sum_{i,j} \frac{D_{ij} n_i n_j}{n(n-1)} \tag{26.2}$$

Suppose there is a point x that belongs to cluster i in the previous iteration. However, in the next iteration, x is closer to cluster j and joins cluster j according to our algorithm. It is not difficult to prove that $d(i+1) < d(i)$, and $D(i+1) > D(i)$.

Now, knowing that we are one step closer to our goal (optimized clustering) after each iteration, we can repeat the process until the most optimized classification is reached.

We can apply the above logic to more generalized machine learning problems. The above algorithm includes two processes and a set of objective functions. The two processes are

1. Starting from the current clustering results, reclassify all data (points). If we consider the ultimate clustering results as a mathematical model, then these cluster centroids (values) and the mapping between points and clusters could be seen as this model's parameters.

2. Obtain new clusters according to the results of reclassification.

The objective functions are the distance between a point and its cluster d and the distance between two clusters D. The whole process aims to minimize the objective function d while maximizing the other function D.

In general, if the observation data (points) are numerous, a computer can learn a model through multiple iterations similar to the two steps of the method above. First, compute the results of all observation data inputted in the current model. This step is called the expectation step, or the E-step. Next, recompute model parameters to maximize expectation. In the above example, we maximize both D and $-d$.* This step is called the maximization step, or M-step. This type of algorithm is called the EM algorithm.

Many algorithms we have introduced in previous chapters are EM algorithms, for example, the hidden Markov model training method, the Baum-Welch algorithm, as well as the maximum entropy model training method—the GIS algorithm. In the Baum-Welch algorithm, the E-step computes the number of transitions between each state (can be a fraction) and the number of times each state produces them according to the current model; the M-step re-estimates the hidden Markov model's parameters according to these numbers. Note that the objective function to be maximized in the training of HMM is the probability of observations. In the maximum entropy model's general iterative algorithm GIS, the E-step computes each feature's mathematical expectation according to the current model, and the M-step adjusts model parameters according to the ratios of mathematical expectations and actual observed values. Here, the objective function to be maximized is the entropy of the probability model.

Lastly, we need to discuss whether the EM algorithm always guarantees a global optimal solution. If our optimized objective function is a convex function, then the EM algorithm can definitely provide a global optimal solution. Fortunately, our entropy function is a convex function. If measured in Euclidean distance in N-dimensional space, the two functions we try to optimize in clustering are also convex functions. However, in many circumstances, including the

FIGURE 26.5: That mountain is even higher, but where is the path?

---

*Maximizing $-d$ is equivalent to minimizing d.

cosine distance in document classification, the functions are not necessarily convex. Hence, it is possible that the EM algorithm provides a local optimal solution rather than a global one (Figure 26.5).

26.3 Summary

After providing some training data and defining a function to be optimized for an EM algorithm, we can leave the rest to the computer. In several iterations, the model we need is trained and ready. This is so beautiful, as if God deliberately arranged it. That is why I call it God's algorithm.

Chapter 27

Logistic regression and web search advertisement

Web search advertisement makes much more money than traditional display ads due to two major advantages that search ads have: knowing what users are looking for according to search queries, and being able to predict which ads the user might click on. As a result, ads inserted in search result pages provide useful commercial information that may help users instead of annoy them.

27.1 The evaluation of web search advertisement

Web search advertisement has evolved through three generations. In the first generation, search ads were displayed according to the bidding prices of advertisers. Simply speaking, whoever gives more money can have their ads displayed first. Baidu and Overture (acquired by Yahoo) were in this generation. Yahoo provided an explanation regarding why this approach makes sense: companies that can afford to pay a lot of money must have good products, so it would not affect user experience. However, this is not the case because those merchants who cut cost and provide low-quality products may have an extremely high profit margin and can spend lots of budget on marketing. In China, the most expensive ads on Baidu.com sometimes are those of counterfeit or pirated drugs. Prioritizing ads only by bidding price will undermine the user experience, and soon users will no longer click on these ads. In the long term, fewer and fewer ads will be clicked. Eventually, advertisers will no longer pay to place ads and the business will shrink.

In fact, this quick money-making method did not bring Yahoo more income than Google. Its revenue per thousand impressions (RPM) was less than half of that of Google. Instead of simply placing the ads of whoever offers more money in top positions, Google predicts which ad is relevant to the user's search intention and maybe to be clicked on, and then determines ad placement according to many factors, such as bidding price, click through rate (CTR), etc. Several years later, Yahoo and Baidu realized that they had a long way to go until reaching Google's level, so they changed their systems accordingly - Panama ranking system for Yahoo ads and Baidu's Phoenix Nest System. These systems are similar to the first generation of Google's system, which we can consider as the second stage of search ads. The key technology here is predicting the probabilities of candidate ads that users may click on, or CTR estimate. The third stage is in fact to achieve global optimization. We will not explain it in detail since it is not relevant to our topic in this chapter.

The best way to estimate CTR is making predictions according to past experience. For example, for a certain search, ad A is displayed 1,000 times and clicked through 18 times, while ad B is displayed 1,200 times and clicked through 30 times. The CTR for these two ads are 1.8% and 2.5%, respectively. Thus, if the two advertisers offer similar prices, it would make more sense to place ad B first.

However, the reality is much more complicated. First, this method would not work for new ads because new ads have no historical CTR data.

Second, even for old ads that have been displayed before, a specific ad corresponding to a search query may be clicked on only two or three times in most cases. Therefore, the statistical data is not conclusive, as we cannot definitively say that the ad clicked on three times is better than the one clicked on two times. In other words, we cannot conclude that the town's gender ratio is 2:3 if we see two men and three women on a certain street.

Third, an ad's CTR is obviously related to where it is displayed - the ad displayed first should have a higher CTR than the ad displayed second. Hence, when estimating CTR, we have to first eliminate this noise. Lastly, we need to point out that there are many other factors that can affect CTR, and we have to consider them when making estimates.

Now, the task has become complicated. It is not easy to describe all of the factors with one unified mathematical model, not to mention that we want the model to be more accurate as the amount of data increases. In the early days, there had been many methods to correct and approximate experience values, but none of them was very effective in integrating all the features. Later on, the industry generally adopted logistic regression, or the logistic model.

27.2   The logistic model

The logistic model fits the probability of an event's occurrence into a logistic curve, the range of which is (0,1). The logistic curve is an S curve whose value changes dramatically in the beginning, gradually slows down, and eventually saturates. A simple logistic regression function has the following form:

$$f_{(z)} = \frac{e^z}{e^z + 1} = \frac{1}{1 + e^{-z}} \tag{27.1}$$

It corresponds to the following curve shown in Figure 27.1.

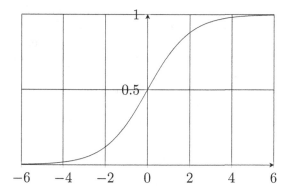

FIGURE 27.1: Logistic regression function curve.

The advantage of logistic regression is that while the independent variable varies between $-\infty$ and $+\infty$, the range of $F(z)$ is between 0 and 1. (Of course, since the value of $F(z)$ basically remains the same when z is beyond $[-6, 6]$, we generally do not consider it outside of that range in practical applications.) We know that the corresponding function between $[0, 1]$ can be a probability function, so the logistic regression function can be linked to a probability distribution. The advantage of independent variable z ranging between $(-\infty, +\infty)$ is that it can combine this kind of signal - no matter how big or small the combined value is, we can always end up with a probability distribution.

Returning to our CTR estimate problem, suppose there are k variables that affect CTR, $x_1, x_2, \ldots, x_k$. Combining them linearly, we will have the following:

$$z = \beta_0 + \beta_1 x_1 + \beta_2 x_2 + \cdots + \beta_k x_k \qquad (27.2)$$

Here, each x_i is a variable. They represent all kinds of information that affect the probability estimates, such as ad positions, relevance between ads and corresponding search queries, display time (for example, ads displayed in the evenings will have a slightly higher CTR than ads displayed in the afternoons). The corresponding β_i is its regression parameter, representing the importance of a variable. β_0 is a special parameter unrelated to any variable. It ensures that there is a stable probability distribution even when there is no information provided.

Let us look at the following example of estimating the CTR of an advertisement related to a search about flowers. Suppose the influencing factors are the numbers of click-throughs per thousand impressions (or the number of impressions required per click), the relevance between the ads and the search queries (variable X_2), the distribution of genders of target population (variable X_3), etc.

Let X_1 be the number of impressions needed per click, X_2 be the relevance between ads and the search queries, which is between 0 and 1 (1 means well-matched while 0 is totally irrelevant), and let X_3 be gender (1 for male, 0 for female). We also assume that the corresponding parameters of X_1, X_2, and X_3 are $\beta_0 = 5.0$, $\beta_1 = 1.1$, $\beta_2 = -10$, $\beta_3 = 1.5$, respectively.

When a male user searches for "flowers" and the ad is about roses, the corresponding variables are $X_1 = 50$, $X_2 = 0.95$, then, $Z = 5 + 1.1 \times 50 + (-10) \times 0.95 + 1.5 \times 1 = 52$, and the estimated CTR is

$$P = \frac{1}{Z} = 0.019 = 1.9\%. \tag{27.3}$$

There are two tricks here. First, we need to know how to select information relevant to the ad's CTR. This is the job of search ads engineers and data mining experts, which we will not describe in detail here. Our main topic is the second point - determination of the parameters.

The logistic function above is in fact a single-layer artificial neural network. If the number of parameters to be learned is not large, then all the methods for training artificial neural networks are suitable here. However, for problems like estimating search ads CTR, there are millions of parameters, which require more efficient training methods. The reader may have already noticed that the logistic function (27.1) and the maximum entropy function that we introduced previously have similar function values and forms. Their training methods are also similar. The IIS method for training maximum entropy models can be used directly for training parameters of logistic functions.

In an advertising system, the quality of the CTR estimation system determines whether the search engine company can increase its RPM. At present, both Google and Tencent's search ad systems have adopted logistic functions in estimating CTR.

27.3 Summary

The logistic regression model is an exponential model that combines the different factors that affect probabilities. It shares similar training methods with many other exponential models (such as the maximum entropy model), and can be implemented through GIS and IIS algorithms. Besides its application in information processing, logistic regression model is also widely used in biostatistics.

Chapter 28

Google Brain and artificial neural networks

In 2016, AlphaGo, the Google computer program that plays the board game Go, beat the world champion, Lee Se-dol. It was the first time a computer Go program has beaten a 9-dan professional without handicaps. This event led to a worldwide rise of interest in artificial intelligence. Several months later, its second version, Master, won all 60 matches against all human Go masters. In 2017, the improved version of AlphaGo beat Ke Jie by 3:0, the world's No.1 ranked player at the time, in a three-game match.

The "intelligence" of AlphaGo comes from its special machine learning algorithm, in particular, a deep learning algorithm.

Google's earliest deep learning algorithms and the corresponding tool "Google Brain" were launched in 2011. It was acknowledged to have a much faster and deeper thinking than other artificial intelligence computers at the time. In order to demonstrate its intelligence, Google provided several examples, e.g., it reduced the speech recognition error rate 16% relatively, from 13.6% to 11.6%. This improvement is very significant because it takes all scientists in the world more than two years to achieve 2% absolute (or 10% relative) error rate reduction. Google achieved this goal not by improving the speech recognition methods, nor by using more training data. It only used a new "brain" to re-train the existing acoustic models' parameters to accomplish this significant error rate reduction.

Now, we have a sense of how intelligent Google Brain is. However, if we were to open up this "brain," we would notice that there is nothing mysterious about it aside from reimplemented artificial neural network training methods using parallel computing technologies. Hence, in order to explain the Google Brain, we need to first talk about artificial neural networks.

28.1 Artificial neural network

Some scientific terms sound full of mystery, and "Artificial Neural Networks (ANN)", or "Neural Networks (NN)" is one of them. When I first heard this concept, I felt quite confused. As we know, the structure of the human brain is too mysterious to comprehend, and an "artificial" neural network sounds like an imitation of the human brain using computers. Considering the complexity of the human brain, it is no surprise that people's first reaction is to think that artificial neural networks are a very profound topic. If you happen to meet a nice and communicative scientist or professor, he may be willing to spend an hour or two to explain artificial neural networks to you in a simple language, and you will say, "oh, I get it now." However, if you unfortunately meet someone who likes to show off his knowledge, he will tell you very seriously, "I am using artificial neural networks" or "my research topic is artificial neural networks," and that will be the end of the conversation. You might respect him for his knowledge and feel inferior about yourself. Of course, it is possible that someone tries to explain the concept to you but uses pedantic terms, and you end up feeling more confused. After having wasted your time and learned nothing, you conclude that you will no longer worry about understanding artificial neural networks for the rest of your life.

Please do not take this situation as a joke. These scenarios have happened to me, personally. First, I did not meet someone who was willing to explain the concept of neural networks to me in an hour or two. Then, I encountered some people who liked to show off. As a young person, I always had the desire to understand what I did not know at first, so I decided to audit a course on neural networks. However, I quit after two or three classes. I did not find it helpful but instead a waste of my time. Since I did not need to use artificial neural networks in my research at the time, I did not bother about it. Later, when I was pursuing my PhD at Johns Hopkins, I had the habit of reading books before bed and by that I mean I read a few textbooks about artificial neural networks. To my surprise, I mastered this knowledge and later was able to use it in several of my projects. Looking back at artificial neural networks, I can tell you that the concepts of this method are not complicated at all, and the entry barriers to learn them are not as high as most people think; I just took some detours.

The word artificial neural network sounds like artificially simulating a human brain. The biology terms like "neurons" make it sound even more mysterious, reminding people of bionics, cognitive science, and other things that are difficult to understand. In fact, besides borrowing some biological terms and metaphors, artificial neural network is completely unrelated to the human brain. It is a directed acyclic graph (DAG) like the ones we have introduced before, albeit a special kind. We already know that directed acyclic graphs include nodes and directed arcs connecting the nodes, and artificial neural networks also have nodes. These nodes are referred to by a new name - neurons. Their directed arcs are viewed as nerves connecting the neurons. Of course, it is a special kind of directed acyclic graph, and its special features can be summarized as follows:

1. As shown in Figure 28.1, all of the nodes are layered; nodes in each layer can point to nodes in the layer one level above through directed arcs, but nodes in the same layer are not connected by arcs. None of the nodes can skip a layer to link to nodes two layers above. Although I only drew three layers of nodes in this figure, the number of layers of an artificial

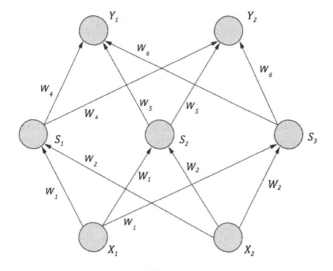

FIGURE 28.1: A typical artificial neural network (three layers).

neural network could be any number in theory. In practical application, most people would not design a network with more than five layers in the past, because the more layers there are, the more complicated it is to calculate.

2. Each arc has a value called the weight. According to these weights, we can use a very simple equation to calculate the values of the nodes they point to. For example, in Figure 28.1, the value of node S_1 depends on the values of X_1 and X_2 (represented with x_1 and x_2, respectively), as well as the corresponding values on the directed arcs w_{11} and w_{21} (we will introduce how to calculate this later).

For ease of communication, some conventional terms are used in literature. For example, the lowest layer of nodes is sometimes called the input layer. This is because in applications, the model's input values are only given to this layer of nodes, and all other node values are obtained directly or indirectly through these input values. For example, in Figure 28.1, the value of S_1 is obtained directly from X_1 and X_2, and the value of Y_1 is obtained indirectly from X_1 and X_2. Corresponding to the input layer at the bottom, nodes in the top-most layer in the graph are called output nodes. This is because the output values we need to obtain through this model are obtained through this layer of nodes. Of course, the other layers in between are the middle layers. Since these layers are not visible to the outside, they are also called hidden layers.

Here, the readers may ask, "is an artificial neural network that simple?" The answer is "yes." Then, they may ask, "can that simple model work?" The answer is also "yes." When I teach in several universities, most students have these two questions in the beginning. However, these simple models have so many applications in fields such as computer science, communication, biostatistics, medicine, finance, and economics (including stock market predictions). All of these areas have problems related to "machine intelligence," and they can be summarized as problems of pattern classification in multidimensional space. The artificial neural network that we described above happens to be very good at pattern classification. We can enumerate the applications of artificial neural networks, including speech recognition, machine translation, face recognition, cancer cell recognition, disease prediction, stock market trends prediction, and so forth.

In order to illustrate how artificial neural networks help us solve the intelligence problems above, we will use a well-known example that we have used in previous chapters - speech recognition. When discussing speech recognition, we mentioned the concept of "acoustic models." In actual speech recognition systems, acoustic models are typically constructed in units of vowels and consonants,* and each vowel and consonant corresponds to a set of data that can be viewed as coordinates in a multidimensional space. Thus, we can map each vowel and consonant to a point in the multidimensional space. In fact, recognizing these phonemes is to segregate this multidimensional space into several regions so that each phoneme belongs to a different region.

In Figure 28.2, we randomly selected the positions of five vowels and consonants a, e, o, t, zh, and the task of pattern classification (speech recognition) is to segregate the multidimensional space and divide the phonemes into different regions.

Returning to our main topic, let us see how an artificial neural network recognizes these phonemes. For the sake of simplicity, suppose the space is only

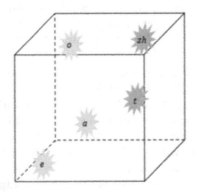

FIGURE 28.2: Vowels a, e, o and consonants t, zh in multidimensional space.

---

*Practical speech recognition systems need to consider each vowel and consonant's correlation with the phonemes before and after it. For the sake of simplicity, suppose each vowel and consonant model is independent.

two-dimensional and there are only two vowels to be distinguished: a and e. Their distribution is shown in Figure 28.3 below.

We have mentioned previously that the task of pattern classification is to draw a line in the space so that those of a and e can be separated. The dotted line in Figure 28.3 is the cutting line, to the left of which is a, and the right side is e. If a new phoneme comes in, it is recognized as a if it falls on the left side, and recognized as e if it falls otherwise.

Now, we can use an artificial neural network to implement this simple divider (the dotted line). The structure of this network is as follows:

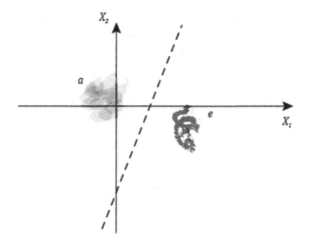

FIGURE 28.3: Pattern distribution of vowels a and e (two-dimensional model).

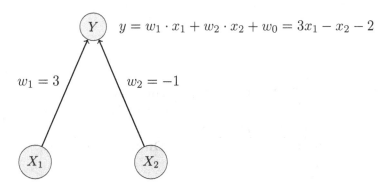

FIGURE 28.4: The artificial neural network that performs pattern classification for the two vowels in Figure 28.3.

This is a very simple artificial neural network. In this network, there are two input nodes: X_1 and X_2, and one output node Y. On the arc connecting X_1 to Y, we assign a weight $w_1 = 3$, and on the arc connecting X_2 to Y, we assign a weight $w_2 = -1$. Then, we set the value of Y as a linear combination of the two input nodes' values x_1 and x_2, i.e., $y = 3x_1 - x_2$. Note that the above function is a linear function, and Y's value can be regarded as the dot product of the input vector (x_1, x_2) and the weight vector (of directed arcs pointing to Y) (w_1, w_2). For our convenience later, let us add a constant term in the equation, -2:

$$y = 3x_1 - x_2 - 2 \tag{28.1}$$

Now, we input the coordinates of some points on the plane $(0.5, 1)$, $(2, 2)$, $(1, -1)$, and $(-1, -1)$ into the two nodes on the first layer (see Figure 28.5), and see what we obtain at the output node (see Table 28.1).

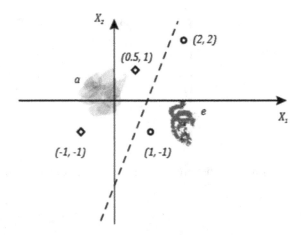

FIGURE 28.5: Pattern distribution of four points using an artificial neural network.

TABLE 28.1: Four different input values of the artificial neural network and their corresponding output values.

Input values (x_1, x_2)	Output value Y
(0.5, 1)	-1.5
(2, 2)	2
(1, -1)	2
(-1, -1)	-4

We can say that if the value we obtain at output node Y is greater than zero, then this point belongs to the first category, e, and otherwise it belongs to the second category, a. Of course, this simple artificial neural network in Figure 28.4 is completely equivalent to the linear classifier in Figure 28.5 $x_2 = 3x_1 - 2$. Thus, we have defined a linear classifier using an artificial neural network.

Of course, we can also turn Y in this artificial neural network into a middle node S, and then add another output layer that includes two nodes Y_1 and Y_2, as shown in Figure 28.6:

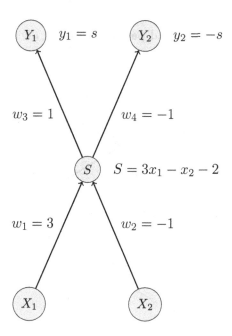

FIGURE 28.6: The artificial neural network after turing Y into the middle node and adding one more output layer.

Thus, the point belongs to the category corresponding to the greater value between output nodes, Y_1 and Y_2.

However, if the distribution of a and e is complicated (such as shown in Figure 28.7), a simple straight line cannot divide them up. Of course, if we can curve the divider, we can still separate the two.

In order to achieve a curved division, we can design an artificial neural network like the one shown in Figure 28.8.

This artificial neural network is not only more complicated in structure (one additional layer), but the computation of each node's value is also more complex. The value of node (neuron) S_2 is calculated using a nonlinear function,

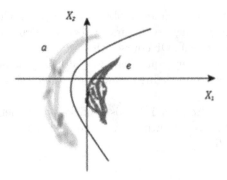

FIGURE 28.7: A complicated pattern has to be divided by a curved line instead of a straight line.

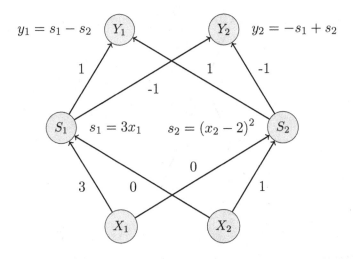

FIGURE 28.8: An artificial neural network designed for the pattern in Figure 28.7 (equivalent to a nonlinear classifier).

called the neuron function. In the example above, calculating the value of S_2 used a square function (a quadratic function). Here, the reader may be wondering: how do we choose what function to use when calculating the value of each node? Of course, if we allow these functions to be randomly selected, the classifier would have a very flexible design but the corresponding artificial neural network would lose generality, and these functions' parameters would be difficult to train. Hence, in artificial neural networks, we require that the neuron function is simply a nonlinear transformation of the result of the linear combination of input variables (the values on its parent nodes). This sounds very complicated, but the following example will help us understand it.

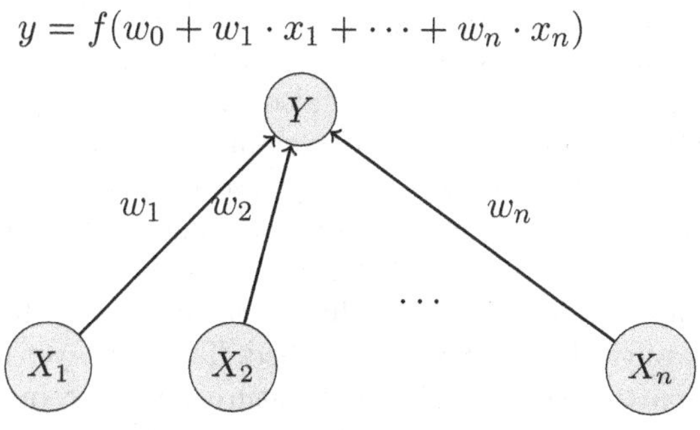

$$y = f(w_0 + w_1 \cdot x_1 + \cdots + w_n \cdot x_n)$$

FIGURE 28.9: The meaning of a neuron function.

Figure 28.9 is part of an artificial neural network. Nodes $X_1, X_2, \ldots, X_n$ point to node Y, and their values are $x_1, x_2, \ldots, x_n$ with respective weights $w_1, w_2, \ldots, w_n$. Computing node Y's value y requires two steps. First, we need to calculate the linear combination of values $x_1, x_2, \ldots, x_n$:

$$G = w_0 + x_1 \quad w_1 + x_1 \quad w_2 + \quad + x_n \quad w_n \tag{28.2}$$

The second step is computing Y's value $y = f(G)$. After the completion of the first step, G already has a fixed value. Although function $f(\cdot)$ itself can be non-linear, it will not be too complicated since it has only one variable. The combination of these two steps ensures the flexibility of artificial neural networks while keeping the neuron functions simple.

In order to further illustrate the rules for neuron functions, we have provided some examples in the Table 28.2 to demonstrate which functions can be used as neuron functions and which cannot.

TABLE 28.2: Some examples and counterexamples of neuron functions.

Function	Neuron function?
$y = log\,(w_0 + x_1 \quad w_1 + x_1 \quad w_2 + \quad + x_n \quad w_n)$	Yes
$y = (w_0 + x_1 \quad w_1 + x_1 \quad w_2 + \quad + x_n \quad w_n)^2$	Yes
$y = w_0 + log\,(x_1) \quad w_1 + log\,(x_2) \quad w_2, \quad ,log\,(x_n) \quad w_n$	No
$y = x_1 \quad x_2, \quad , x_n$	No

Returning to the artificial neural network shown in Figure 28.8, we have obtained the following classifier:

$$3x = (y - 2)^2 + 2 \tag{28.3}$$

This classification has a curved border. Theoretically, as long as the artificial neural network is well designed, any complicated curved border (or a curved surface in a high-dimensional space) can be implemented. Now that we have some basic understanding of artificial neural networks, let us summarize what we have covered so far.

An artificial neural network is a layered directed acyclic graph. The first layer with input nodes $X_1, X_2, \ldots, X_n$ (the bottom layer in Figure 28.1) receive input information. It is also called the input layer. Values from these nodes $(x_1, x_2, \ldots, x_n)$ are weighted linearly according to the weights of their output arcs $(w_0, w_1, w_2, \ldots, w_n)$ to obtain G, and it undergoes a function transformation $f(G)$ before being assigned to the second layer's node Y.

The second layer nodes propagate this value accordingly to the third layer, and this goes on until the last layer is reached. The last layer is also called the output layer. In pattern classification, the feature (such as coordinates) of a pattern (image, speech, text, etc.) propagates layer by layer according to the rules and equations above, starting from the input layer. Eventually, at the output layer, the inputted pattern is classified into the category whose corresponding node has the greatest value.

These are the basic principles of artificial neural networks.

In an artificial neural network, there are only two parts that need to be designed. The first one is its structure, i.e., the number of layers, the number of nodes in each layer, and how nodes are connected in the network; the second part is the nonlinear function. A commonly used function is the exponential function, i.e., $f(\cdot)$:

$$f(G) = e^G = e^{w_0 + x_1\, w_1 + x_1\, w_2 + \ \ + x_n\, w_n} \tag{28.4}$$

Here, its pattern classification ability is equivalent to that of a maximum entropy model.

It is worth noting that if we consider the values of different output nodes as a kind of probability distribution, then an artificial neural network is equivalent to a probabilistic model such as the statistical language model that we have mentioned previously.

We have yet to discuss how the weights on the arcs, which are the model's parameters $(w_0, w_1, w_2, \ldots, w_n)$, are obtained. Similar to many machine learning models, they are obtained through training (learning). This is our topic for the next section.

28.2   Training an artificial neural network

Similar to many machine learning algorithms that we have mentioned previously, artificial neural network's training is divided into supervised training and unsupervised training. Let us first look at supervised training.

Before training, we need to first obtain some marked samples (training data), like the data in Table 28.1 with input data x_1, x_2 and their corresponding output value y. The objective of training is to find a set of parameters w (weights) so that the model's output value (a function of the parameters, notated as $y(w)$) and the pre-designed output value of this set of training data (suppose it is y) are as close as possible.* We can state the above ideas formally as the following:

Suppose C is a cost function. It represents the difference between the output values (classification results) according to the artificial neural network and the output values of actual training data. For example, we can define $C = \sum (y(w) - y)^2$ (i.e., the Euclidean distance). Our training objective is to find parameter $\hat{w}$ so that

$$\hat{w} = \arg \min_{w} \sum [y(w) - y]^2 \tag{28.5}$$

Now, the artificial neural network training problem has become an optimization problem. Simply speaking, it is the same as finding the maximum (or minimum) value in high school math. A commonly used method to solve optimization problems is the gradient descent method. We will not introduce this algorithm in detail as it can be found in many reference books, but we can illustrate its principles with an example. We can think of the process of finding the maximum value as mountain climbing.† When we climb to the top of the mountain, we have found the maximum value (of course, descending to the bottom of the valley is finding the minimum value). Now, how do we hike to the top with the fastest speed? According to the gradient descent method, we always take a step toward the "steepest" direction to ensure that we hike to the top in the shortest amount of time.

After obtaining the training data, defining a cost function C, and finding the set of parameters that minimize the cost using the gradient descent method, we have everything ready. The training of an artificial neural network is complete. However, in real applications, we usually cannot obtain enough annotated data. Therefore, in most circumstances, we have to train the artificial neural network's parameters by unsupervised training.

Unlike supervised training, unsupervised training only has input data (x) without corresponding output data (y). Thus, without knowing the difference between the output values produced by the model and the actual output values, the above cost function C can no longer be used. Hence, we need to define a new (and easily computed)‡ cost function that can judge (or estimate) the quality of a trained model without knowing the correct output values. Designing such a cost function is a challenge in itself. Artificial neural network researchers need to find a suitable function according to the specific application. However, generally speaking, cost functions have to follow this principle: since artificial neural networks solve classification problems, we expect that after classification, samples (training data) in the same category are close to each other, while those in

---

*Here, we say "as close as possible" because given a certain artificial neural network structure, there does not exist a parameter combination so that all training data are completely identical to the model's output. This is a common phenomenon in machine learning.

†Finding minimum problems is similar.

‡Sometimes the computation of a cost function can be highly complicated.

different categories should be as far apart as possible. For example, for the multidimensional space pattern classification problem mentioned previously, we can make the average Euclidean distance from sample points to the trained centroid as the cost function. As for estimating the conditional probabilities of language models, we can use entropy as the cost function. After defining the cost function, we can proceed with unsupervised parameter training using the gradient descent method.

We need to point out that for structurally complex artificial neural networks, the amount of computation required for training is very large, and this is a NP-complete problem. Therefore, many machine learning experts are looking for good approximation methods.

28.3   The relationship between artificial neural networks and Bayesian networks

We can see from the above descriptions that artificial neural networks and Bayesian networks look alike in many aspects. For example, the directed acyclic graph in Figure 28.8 can also be regarded as a Bayesian network. Actually, the two types of networks are common in the following aspects:

1. They are both directed acyclic graphs. The value of each node only depends on the nodes of the preceding level, independent of nodes of two levels beyond or more. In other words, they abide by the Markov hypothesis.

2. Their training methods are similar. We can see this from the above description.

3. For many pattern classification problems, these two methods achieve similar results. In other words, many problems that can be solved using artificial neural networks can also be solved with Bayesian networks, and vice versa. However, one may be more efficient than the other. If we were to consider artificial neural networks and Bayesian networks as statistical models, then the two models have similar accuracies.

4. Their training processes both require enormous amounts of computation. Therefore, we should be well prepared when using artificial neural networks.

Nonetheless, artificial neural networks and Bayesian networks also have quite a few differences:

1. In terms of structure, artificial neural networks are completely standardized, whereas Bayesian networks are more flexible. Google Brain chose to use artificial neural networks because of this feature.

2. Although neuron functions in neural networks are nonlinear, the variables have to be linearly combined first and only one variable (the result of the combinations) will be transformed nonlinearly. Thus, neuron functions are easy to implement with a computer. However, in a Bayesian network, variables can combine into any function without limitations. With added flexibility, the complexity also increases.

3. Bayesian networks can use contextual relevance more flexibly and are thus often used to decode an input sequence, such as recognizing speech into text or translating an English sentence into Chinese. In contrast, the output of artificial neural networks is rather isolated. It can recognize words individually but has a harder time processing a sequence. Hence, it is used to estimate parameters of a probabilistic model, such as acoustic model parameter training in speech recognition and language model parameter training in machine translation, and so on.

If you know the similarities and differences between artificial neural networks and Bayesian networks, you will notice that many mathematical tools of machine learning are related, and we can find the most convenient method for a specific real problem.

28.4 Extended reading: "Google Brain"

Suggested background knowledge: algorithms, numerical analysis.

In simple words, Google Brain is a large-scale parallel-processing artificial neural network. Aside from its size, what are its other advantages compared to a regular artificial neural network? Theoretically, there are none. However, the "large scale" itself can make a seemingly simple tool very powerful. In order to illustrate this, we need to demonstrate the limitation of "small scale."

Here, we have to discuss the historical development of artificial neural networks. The concept of artificial neural networks originated in the 1940s, and it was implemented on computers in the 1950s. People were already able to use artificial neural networks to solve simple classification problems at that time. Although many people speculated whether such a simple model might eventually "possess intelligence," this was not true. By the end of the 1960s, the famous artificial intelligence expert, Marvin Minsky, found two reasons for this:

First, an (overly simplistic) artificial neural network with only input and output layers (no middle layers) cannot accomplish even the simplest exclusive or operations (because they are not linearly separable).

Second, for slightly more complex artificial neural networks, the amount of computation needed to train them exceeded the capacity of computers at that time, making it impossible to implement them at that time.

In fact, these two reasons can be combined into one. If the scale of artificial neural networks is too small, they cannot solve any sophisticated problem; if it is too large, then they cannot be implemented in computers. Hence, from the 1970s to the 1980s, artificial neural networks were neglected. In early 1990s, the speed and capacity of computers increased exponentially, due to Moore's Law. The computational power of computers had increased by tens of thousands of folds from the 1960s till then. So, scientists and engineers could finally train larger-scale artificial neural networks. Under such circumstance, artificial neural networks could be used to recognize handwriting or perform small-scaled speech recognition. However, artificial neural networks were still unable to solve bigger problems, so this technology was neglected again after a hot period of a few years. This shows that the scale of an artificial neural network determines its power and potential applications.

After 2010, Moore's Law increased the computational power of com- puters by another tens of thousands of folds (compared to the 1990s). At the same time, the rise of cloud computing made it possible to use tens of thousands of processors simultaneously. Thus, artificial neural networks were able to accomplish even greater tasks. However, this time the increase of computational power mainly came from parallel computing, instead of that from the speed of a single processor two decades before. That means we can simply increase the number of processors by thousands or millions of folds to accomplish a big task. Hence, the old training methods for artificial neural networks had to be modified to fit the needs of parallel computing. Google Brain was invented under such circumstances, and its innovation was to use millions of processors in cloud computing data centers in parallel. Why did Google Brain adopt artificial neural networks instead of other machine learning technologies? There are three reasons for this:

First, theoretically speaking, an artificial neural network can "draw" pattern classification borders of any shape in a multidimensional space, which makes it highly versatile.

Second, although many machine learning algorithms emerged and improved in the past twenty years, the artificial neural network algorithm stayed stable with few changes. Google hoped that its computing toolkits (Google Brain) could be used in many applications for a long time once it has been implemented. If the toolkits were based upon a machine learning algorithm that had to be updated every year, the basic structure and user interfaces also would have to change. Therefore, all the development based on the toolkits would have to be re-implemented as well. As a result, no third party developers will use such toolkits.

Third, not all machine learning algorithms (such as Bayesian networks) are suitable for parallel processing. The artificial neural network training algorithm is relatively simple and easy to be implemented in parallel.

Now, let us look at how Google Brain is implemented. Its training algorithm shares many similarities with the design ideas of MapReduce. Both of them use the divide-and-conquer method. Their difference is that the divide-and-conquer algorithm for Google Brain is more complicated. In order to illustrate this, let us look at this five-layer artificial neural network in Figure 28.10.

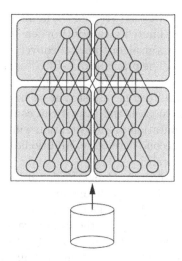

FIGURE 28.10: Training an artificial neural network in four chunks.

In order to make the graph look clearer, many links between nodes have been ignored. In this artificial neural network, the time complexity of its training process is very high. Hence, we divide this network into four chunks as shown above. Of course, in Google Brain, an artificial neural network may be cut into thousands of chunks. Unlike in MapReduce, the computation of each chunk is not completely independent but needs to take many adjacent chunks into consideration. When the number of chunks is large, the total number of associations is roughly proportional to the square of number of chunks. Although this would make the computation between chunks rather complicated, it breaks down a large problem that could not be solved on one server into a number of small problems that could be solved on one server.

In addition to its ability to train artificial neural network parameters in parallel, Google Brain made two improvements to reduce the amount of computation.

First, it reduced the amount of computation of each iteration. Google Brain used the stochastic gradient descent algorithm instead of the gradient descent method that we mentioned previously. This algorithm does not need to compute all samples like the gradient descent method, but it randomly selects a small amount of data to compute the cost function. Thus, it reduces the computation complexity significantly at the cost of a slight loss in accuracy. Because Google Brain has an enormous amount of training data, it would take too much time using the traditional gradient descent method for each iteration. Therefore, the faster stochastic gradient descent method that can balance between computation time and accuracy is used. The stochastic gradient descent method is a faster algorithm that strikes a good balance between computation time and accuracy.

The second improvement was to reduce the number of iterations in training. Google Brain adopted the limited-memory Broyden-Fletcher-Goldfarb-Shanno method (L-BFGS), which is an algorithm that converges sooner than the regular gradient descent method. The principle of L-BFGS is similar to that of the stochastic gradient descent algorithm, but it is slightly more complex. Its advantage is that the step length can be adjusted at each iteration according to the "distance" toward the final goal, and thus it can converge in fewer iterations. However, the total computation of each iteration will increase slightly because it needs to calculate the second derivative. Additionally, the L-BFGS method is easier to implement in parallel.

With these two improvements, Google Brain can accomplish artificial neural network training tasks with overwhelming amounts of computation.

Next, let us look at Google Brain's storage problem. Since only the input ends can access with the training data, these data are stored locally on the input server (calculation module) (see Figure 28.11). On the other hand, the model parameters obtained through each iteration's training on every server need to be collected, and then sent to the corresponding calculation module server before the next iteration starts. Hence, these model parameters are stored separately on another set of servers, as shown in Figure 28.12.

Now, we have introduced the design principles of Google Brain. Google Brain's algorithm is summarized as follows:

1. Define two services: Fetching parameters and pushing back parameters.

2. For the nth server, repeat the following steps:
   [/ / The loop begins

3. Fetch parameters and read training data.

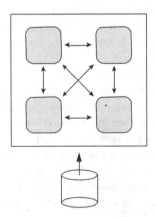

FIGURE 28.11: After breaking down a large artificial neural network training problem, each subproblem is only relevant to the surrounding problems.

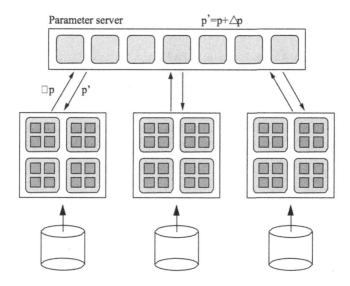

FIGURE 28.12: Google Brain data transmission diagram.

4. Compute the cost function's gradient.

5. Compute the step length.

6. Compute new parameters.

7. Push back new parameters
   ] / / The loop ends

28.5 Summary

Artificial neural networks are machine learning tools with a simple form but a powerful classification ability. We can see the beauty of mathematics once again from artificial neural networks. In practice, all versatile tools must be simple in their form.

Artificial neural networks are closely related to other machine learning tools mathematically, which also shows the beauty of mathematics. This allows us to learn other machine learning methods once we understand neural networks.

Google Brain is not a brain that is capable of thinking any thoughts, but an artificial neural network with a very powerful computational capacity. Hence, rather than saying that Google Brain is very intelligent, it is more accurate to say that it is great at computation. Nonetheless, from another perspective, computation-intensive but simple methods sometimes can solve very complicated problems as the computation power continues to increase.

Bibliography

1. Quoc V. Le, Marc Aurelio Ranzato, Rajat Monga, Matthieu Devin, Kai Chen, Greg S. Corrado, Jeff Dean, Andrew Y. Ng. "Building High-level Features Using Large Scale Unsupervised Learning." Proceedings of the 29th International Conference on Machine Learning, 2012.

2. Andrea Frome, Greg S. Corrado, Jonathon Shlens, Samy Bengio, Jeffrey Dean, Marc Aurelio Ranzato, Tomas Mikolov. "DeViSE: A Deep Visual-Semantic Embedding Model." 2013.

Chapter 29

The power of big data

In Chapter 19, we discussed the importance of mathematical models. Recall that we mentioned the success of astronomer, Johannes Kepler, was largely due to the huge amount of observation data gathered by his teacher, Tycho Brahe. Similar to models, data is also crucial, but people ignored its importance for a long time. If someone proposed a new model, an effective algorithm or a new approach, he would easily gain respect from his peers. A new model is often seen as a significant breakthrough, or even a epoch-making contribution (depending on the specific contribution). However, improvements from gathering and processing a huge amount of data were less valued as people thought that those works lacked creativity. Even just a decade ago, it was comparatively difficult for researchers who use data-driven methods, or empirical methods to publish papers than those who do purely theoretical research. Overall, most people in academia overemphasized the importance of new approaches while underestimating the role of data.

However, with the development of the internet, especially the rise and popularization of cloud computing, this situation has changed significantly. Because of the rapid increase in computers' data gathering, storing, and processing capacities, people began to discover theories and laws from large amounts of data that were hard to find in the past. Hence, many scientific research and engineering fields, e.g., speech recognition, natural language processing, and machine translation, as well as other research areas related to information technologies, such as bioinformatics, biomedicine, and public health, made unprecedented progress. More and more people realized the importance of data. Starting from 2005, a new concept emerged among media—"big data."

What is big data? Is it another concept hype? Is it the same thing as "vast amounts of data" that people used to call it? If not, how are the two concepts related? When speaking of big data, people will naturally think of these questions, and we will answer them in this chapter. However, before answering these questions, we will first discuss the importance of data, then demonstrate the miracles that have happened and will happen using big data. I believe that you will have answers to the questions above after reading this chapter.

29.1 The importance of data

We can say that data accompany us throughout our lives. Then what is data? Many people think that data is just some numbers. For example, experimental data and statistical data that we commonly speak of are presented in the form of numbers. However, these are only data in the narrow sense. Some people think that data can also include information and intelligence. For instance, a word that we often mention, database, is the amalgamation of data that follow a certain format. The data in a database could be the basic information about people in an organization, such as name, gender, age, contact information, education, and resume. Most of these data are in the number form. Some people also think that the definition of data can be even broader to include information of any format, such as content on the internet, archival materials, design blueprints, medical records, videos and images, etc. This is data from a broad view. In this chapter, we will discuss data from a broad view. It includes everything we have mentioned above.

To some extent, human civilization and progress are accomplished through gathering, processing, and utilizing data. In prehistorical times, our ancestors had already been using data before tools for recording information were invented. In ancient Chinese legends, there is the story of Fu Xi inventing the Bagua (see Figure 29.1). Fu Xi was one of the three prehistorical emperors in Chinese legends who reigned earlier than the Yan and Huang emperors. Some people say that Fu Xi was in fact a tribe instead of a person, but this is not important. According to the ancient Chinese mythology, Bagua or the eight trigrams he invented could be used to predict the future.

We will not comment on whether Fu Xi's Bagua actually work or not, but this story demonstrated that in prehistorical times, people already knew to use different conditions (which are in fact input data) to classify good and bad lucks in the future into 8 or 64 categories (output data). How could they categorize the future like this and convince so many people? (Although I am personally skeptical.) This is because many people believe that what they have heard and seen in the past (also data) can prove the accuracy of this kind of categorization. For example, if the weather is bad when a troop is preparing for war, the likelihood of winning the war will be relatively low. Knowledge of these events (data) will be passed on to offsprings, and over the generations, people will begin to associate the weather (from heaven) with the results of war and describe them abstractly in the form of trigrams. By the time of agricultural civilization, many ancestral life experiences such as when to sow and harvest plants are often associated with the position data of stars and planets in the sky. Back then, there were no written systems or seldom could people read and write, so this knowledge had to be passed down from one generation to the next in oral languages. There are similar examples in the Western world. For instance, in the Bible, Joseph predicted that seven years of great plenty would be followed by seven years of famine. This is in fact a rough summary of cycle of agriculture production based on climate cycle or another periodical facts, based on experiences (data) of people at that time.

After the Renaissance, modern natural sciences began to emerge and develop rapidly. Regardless of the field, scientists conducted experiments as a very important task in scientific inquiry. The objective of experiments is gathering

FIGURE 29.1: Bagua.

data, which were used by scientists to discover new laws of the universe. Many famous scientists in the world, such as Galileo Galilei, Tycho Brahe, Antoine Lavoisier, and Pierre and Marie Curie, etc. spent their lifetimes conducting experiments and gathering data. In the age of Galileo Galilei, the Chinese pharmacologist and herbalist, Li Shizhen, wrote the Chinese medicine bible *Compendium of Materia Medica*, in which all Chinese medicines were classified and analysed. However, until the popularization of the internet, the global data volume was "very small" by today's standard. This might be one of the reasons that many people overlooked the importance of data in the past.

The importance of data is not only apparent in scientific research, but also permeates every aspect of social life. At Google, all the product managers follow this rule: without data, no conclusion can be made. This is because many intuitive feelings will lead us to opposite conclusions that data leads us. Without using data, the probability of our success will be much lower. In order to illustrate this, let us look at a few examples to understand how large the difference is between our imagined conclusion and the actual data.

The first example is about basic facts.

What are the 10 most populated cities in the world? I presented this question to a dozen people in 2012, and their answers were cities like this: Shanghai, Chongqing, Tokyo, Beijing, and Mumbai. These are metropolises in countries with a lot of people, like China and India, or famous cities like Tokyo, New York, and Paris. In fact, besides Shanghai, New York, Tokyo, and Delhi, 6 out of the 10 cities with the world's highest population densities are what most people would not expect. WorldAtlas summarized the 2012 census results of countries around the world* with the most populous cities as follows in Table 29.1:

---

*http://www.worldatlas.com/citypops.htm#.UaZ9m5WjJFI

TABLE 29.1:   Ten cities with the largest populations in the world (as of 2012).

Ranking	City	Country	Population
1	Tokyo	Japan	37,126,000
2	Jakarta	Indonesia	26,063,000
3	Seoul	South Korea	22,547,000
4	Delhi	India	22,242,000
5	Shanghai	China	20,860,000
6	Manila	Philippines	20,767,000
7	Karachi	Pakistan	20,711,000
8	New York City	United States of America	20,464,000
9	São Paulo	Brazil	20,186,000
10	Mexico City	Mexico	19,463,000

Without looking at the data, people might not expect Seoul, Manila, Karachi, São Paulo, or Jakarta to be on the list.

The second example can demonstrate how far off from the truth we can be when estimating things we do not know.

Dr. Yan Weipeng, the former general manager of Tencent Search Ads, once posed a question: for a 3cm × 5cm gaming advertisement on the front page of a major Chinese portal website (Sina, Tencent, Sohu, or NetEase), how much is the cost on average to obtain a clickthrough? At the time, many people guessed 10 CNY,* 30 CNY, or 50 CNY. I had some experience in advertising, and made a bold guess of 100 CNY. According to Dr. Yan, the cost is in fact over 1,000 CNY (because the clickthrough rate is extremely low, less than 0.01%, even including accidental clicks). Hence, an impression ad is practically useless. Such data is very useful for a company's marketing department to know.

The third example shows that before seeing the data, people often tend to overestimate themselves or overemphasize something's positive effects and overlook its drawbacks.

The return of the US stock market is about 7.5% annually in the last 20 years (1997-2017). About 30% to 50% of middle income people trade stocks by themselves, and this ratio is even higher in males. However, statistics show that 95% of individual investors underperform the market, and 50% to 70% of individual day traders even lose money. Among my highly educated and highly intelligent colleagues and friends, few individuals outperform the market, even though they have master's and Ph.D. degrees from top universities. When I ask them why they still do it despite the low return, besides some who say it is just for fun, most people believed they could do better than professional investors and beat the market.

Most people are overconfident about their ability to make money in the stock market, since it looks so easy. However, the statistics data show completely

---

*1 CNY (Chinese Yuan)=0.15 USD (US Dollar).

opposite results—it is almost guaranteed that individuals underperform the market, even some lucky guys may earn big money from the market sometimes. This example shows us that decisions not supported by data are often inaccurate: rare, individual success cases are amplified in people's impressions, but the risk is minimized. This example also reflects the difference between individual data (individual cases) and large amounts of data.

Next, I asked how many people believe that funds managed by professional investors would bring them better return than the market index, such as the SP 500 index. Many believed this. However, the fact is that 70% (sometimes 90%) of funds underperform the index in one year and seldom can any mutual fund beat that index in the long term. People might be surprised to learn this result, but this is the truth. This example tells us how far apart our imagination is from the fact. Before obtaining enough data, it is difficult for us to make correct judgments. By the way, some readers might ask, if neither individual investors nor funds perform better than the market, then where does all the money go? This answer is very simple. Transaction fees and all sorts of taxes (such as stamp duty, US stock market investment income tax, etc.) eat up a large chunk of the return, and the fund management fees eat up another large chunk. For a dynamically managed fund, if the annual managing fee is 2% (which is common), it would eat up about half of the return in 30-40 years though the 2% rate does not seem high.* To some extent, the stock market is a zero-sum game. The salaries of Securities and Futures Commission officials and fund managers as well as their luxurious working conditions all come from money in the market. The luxury cars and masons of fund managers are also bought with investors' money. Hence, if an individual investor really abides by what the data says, the only sound investment policy would be to buy index funds. Of course, this is not my original idea, but that of famous economists like William F. Sharpe and Burton G. Malkiel.

I told all these stories only to emphasize one point: data is not only important in scientific research but crucial in every aspect of life. It should become the basis of our everyday decisions.

29.2   Statistics and information technology

After obtaining data, we need to employ a scientific tool - statistics - to use data smartly. Today, probability and statistics are taught in one course for non-math majors in many universities, but they are two separate academic disciplines despite the fact that they are closely correlated. Probability is a branch of mathematics that studies quantitative laws of random phenomena. Statistics

---

*For example, if the managing fee and the transaction cost make up 2% of the principal, then an 8% annual return will be reduced to only 6%. This reduction does not seem to be very much, but the return will be reduced in half over 35 years. This is because at the rate of 8%, the total return over 35 years is 1278%, and the actual return over the same period is only 668%.

infers the nature of the target object through collecting, processing, and analyzing data (among other means); it can even be said that statistics is a general science that predicts the future of the target object. We have introduced many applications of probability theory in information technology, but those probabilistic models were often obtained through the statistics of data.

Trustworthy statistics first requires an adequate amount of data. In Chapter 3, "Statistical Language Model", we described that estimating all parameters of a language model requires a "significant amount" of training data in order to obtain a reliable probabilistic model. Why does the data need to be a "significant amount," or "enough"? We will show the reason by the following example.

If you count the total number of people passing through the front gate of the Johns Hopkins University in one hour, and see 543 men and 386 women, you can roughly obtain the conclusion that "this university has slightly more men than women." Of course, you cannot say that the men to women ratio is 543:386, because you know that the statistic can be random and have errors. You can only draw a conclusion like "the ratio of men to women is roughly. 6:4." As long as the rate of men and women passing through the gate follows this pattern, no one will challenge your conclusion after you have counted over 900 samples. However, the situation would be different if you woke up in the early morning and stood at the gate for two minutes, saw three women and one man there, and concluded that three quarters of the students at this university were women (see Figure 29.2). This conclusion obviously will not be acceptable because such a size of sample data is too small, and these cases can be a completely random incident. On a different day or at a different time of the day, you might notice that all of the four people passing through the gate in two minutes are men. If so, you cannot make the conclusion that "this university only enrolls men" either. I think most readers will agree that if the number of samples is too small, then the statistic is completely meaningless. As for how much data is needed to obtain accurate statistical results (in our problem here, it is estimating probabilities), we will need to perform a quantitative analysis.

A hundred years ago, Russian mathematician, Pafnuty Lvovich Chebyshev (1821-1894), proposed an inequality, the Chebyshev Inequality, to show why we need a significant amount of data before we can draw a reliable conclusion from statistics:

$$P(|X - E(X)| \geq \epsilon) < \frac{\sigma^2}{n\epsilon^2} \tag{29.1}$$

Here, X is a random variable, $E(X)$ is the mathematical expectation of this variable, n is the number of experiments (or the number of samples), ϵ is the standard deviation, and σ is the variance. This inequality means that when there are enough samples, the difference between a random variable (such as the gender ratio of people passing through the university gate) and its mathematical expectation (such as the gender ratio of students at the university) can be very small.

FIGURE 29.2: We cannot predict the gender ratio of people in Oxford University from this picture, as there are only eight samples.

Applying the Chebyshev Inequality to our example here, suppose this university's men to women ratio is roughly 6:4, then how many samples do we need to collect to obtain an accurate estimate with a standard deviation smaller than 5% (i.e., confidence 95%)? Inferring from Equation (29.1), we know that this number is about 800, which means that one needs to observe 800 or more pedestrians passing through the gate. Similarly, if we were to make an accurate estimate of the parameter of a bigram model in the Chinese language P(weather| Beijing), suppose this conditional probability is about 1%, then we need to observe the set "Beijing weather" appear about 50 times, which means that "Beijing" has to appear at least 5,000 times. Suppose the probability of "Beijing" appearing in a text is about one thousandth, then we need at least 5 million words to estimate the conditional probability of P(weather|Beijing) precisely. We know that both of these words are common words that appear in samples frequently. As for those uncommon words, in order to gather samples in which they appear enough times, we will need an enormous corpus. In information processing, any problem that involves probability requires a lot of data to support. Hence, we can say that data is our raw material for data processing.

Besides a large amount of data, a representative sample is also required for accurate statistics. Sometimes a large amount of data does not necessarily guarantee accurate statistical results. The data used for statistics must accord with the target task. In order to demonstrate this point, let us look at a case where a large amount of data did not lead to an accurate statistical estimate.

Prior to the presidential election of 1936, *The Literary Digest* predicted that the Republican candidate, Alfred Landon, would win. Before that, *The Literary Digest* had already successfully predicted the results of the previous four consecutive presidential elections. This time, the magazine collected 2.4 million surveys,* which was a lot more than the previous times. Such a large amount of data should be enough to predict the result, and therefore believe their prediction. However, a little-known journalist (also a statistician) called George Gallup made the opposite prediction of the election result. From the statistics of only 50,000 samples, he concluded that the Democrat candidate, F.D. Roosevelt, would stay in office. After the result came out, it turned out that Gallup, who used much fewer samples, was right. Dr. Gallop later explained the reason for this to the confused public: although *The Literary Digest* collected much more samples, their samples were not representative of the American public. The magazine's survey distributors sent out surveys according to the addresses of magazine subscribers, car owners, and those in the Yellow Pages. However, only half of the American families had telephones at home, and even fewer families owned cars at that time. These families usually had higher income, and most of them supported the Republican Party. In contrast, when Gallup was designing his statistical samples, he considered the electorate's race, gender, age, income, and other factors. Hence, even though he only collected 50,000 samples, they were more representative of the American public. This example shows us the importance of representative statistical samples.

However, it is not easy to design representative samples. The story did not end there. The 1936 presidential election prediction not only made Gallup famous, but gave birth to the most authoritative poll company today - Gallup, Inc. After that, this company successfully predicted the election results in 1940 and 1944. Before the presidential election in 1948, Gallup published a conclusion that it thought was highly accurate - the Republican candidate, Thomas Dewey, would defeat the incumbent president and Democrat candidate, Harry S. Truman, by a large margin. Since Gallup made successful predictions three consecutive times, many people believed the prediction this time. However, as we all know by now, Truman ended up winning by a large margin. This not only took many people by surprise, but also made people question Gallup's poll methods - although Gallup considered the electorate's income, gender, race, age, etc., there were many other factors and the combination of such factors that Gallup had failed to consider.

Of course, if the sample data is representative and sufficient, then the statistical results from such data can provide a significantly meaningful guidance to our work, and they are very helpful for improving product quality. The competition in the IT industry today is already a competition of data to some extent.

Let us first look at the competition in the field of web search. Most people think that Google search is slightly better made (in terms of quality) than Microsoft's Bing because Google uses a better algorithm. This view was of course right before 2010, because Bing search was behind Google in technology and engineering back then. However, these two companies are already comparable in terms of technology. Besides its slightly better product design, Google's

---

*Almost ten million surveys were sent out.

advantage largely relies on the power of data. Today's search engines are very different from those before 2000. Back then, the search algorithms were not effective, and one improvement in the algorithm could bring significant progress like a 5% or higher increase in search accuracy. However, there is no more magic method today that could increase the accuracy rate by even 1%. All of the search engines today, including Google, Bing, and China's Baidu, have similar results for common searches. The differences among the search engines are only apparent for those uncommon long tail queries, and the key to improving these search qualities is largely depending on the size of data. Hence, data is the number one factor that determines the quality of a search engine, even more important than algorithms.

Among all sorts of data used in searching, there are two kinds of important data: the webpage data and the user clickthrough data. If the search engine is good, it has to have good webpage data. In other words, the index size must be large, and the content must be up-to-date. This is not too hard to understand, because you cannot make bricks without straw after all. From the perspective of engineering, this uses up a lot of money - as long as you have enough engineers and servers, it is not hard to achieve. However, webpage data alone is not enough. We also need a large amount of clickthrough data. In other words, we need to know that for different search keywords, which search results (webpages) most users click on. For example, for the search "robot," page A is clicked on 21,000 times, page B is clicked on 5,000 times, page C is clicked on 1,000 times, ... According to these clickthrough data, we can train a statistical model to decide the order of search results (i.e., sort the pages in the order of A, B, C, ...). The search algorithm of this model is called the "click model" - as long as there is enough data, this kind of clickthrough-based search sorting is highly accurate. The click model contributes at least 60% to 80%* of weights in the ranking of the search results. That is to say that all of the factors in search algorithm added together are not as important as the clickthrough model.

In order to obtain the clickthrough data, as soon as the search engine comes online, the company collects the users' clickthrough information through logging each search. Unfortunately, the accumulation of clickthrough data is a long process, and it cannot be accomplished in the short-term using money like downloading webpages. For those uncommon searches (commonly known as long tail searches), such as "Picasso's early works," collecting "enough data" to train a model requires a very long time. Considering certain search results' timeliness, search engines with low market shares could not gather enough statistical data before the popular search loses its popularity, so they end up with inaccurate clickthrough models. This is the main reason that Microsoft's search engine lagged behind Google's for a very long time. Similarly, in China, Sogou, Soso, and Youdao's market shares are minimal compared to that of Baidu, so

---

*Although different companies' search engines rely on the clickthrough model to different extents, the weight is at least 60% for all of them.

they have a very hard time training effective clickthrough models. Thus, we can see the Matthew Effect in the search industry, i.e., the search engines with low search volume have lower search qualities because of insufficient user click-through data, and high-quality search engines become even better because of their large amounts of data.

Of course, companies that enter the search market later will not resign to lose. They can use other approaches to quickly obtain data. The first method is to buy traffic. After Microsoft took over Yahoo's search, its search volume rose steeply from 10% of Google's traffic to about 20%-30%. The ranking esti-mated by the clickthrough model became more accurate, and the search quality quickly improved. However, doing this alone is not enough. Hence, some compa-nies came up with more aggressive approaches, such as gathering user click-through behaviors through the toolbar, browser, or even the input method. The advantages of these approaches are that they not only gather the click-through data when the user is using the search engine itself, but also collect data when the user is using other search engines. For example, Microsoft col-lected the clickthrough data when the user is using Google search on Internet Explorer. Thus, if a company has a large market share in the browser market, it can collect large amounts of data even though its search volume may be small. With these data, especially the clickthrough data of users using better search engines, a search engine company can quickly improve the quality of long tail searches. Of course, some people criticize Bing for "copying" Google. In fact, this is not direct copying, but using Google's results to improve its own click-through model. The situation is the same with the Chinese market. Therefore, the competition of search quality became the competition of market shares in browser or other client software markets.

Of course, data is not only helpful to improve search quality, but it is also useful in many aspects in information processing. We can look at two examples: how Google uses large amounts of data to improve the quality of machine trans-lation, and how it improved the quality of speech recognition.

In 2005, Google surprised academic people in the natural language process-ing area by its Rosetta machine translation system. Google, with less than two-year experience in machine translation and being little-known by this society, suddenly became the far number one player in this research area. Even though people knew that the company hired the world-renowned machine translation expert, Franz Och, nobody thought that Google could develop a decent system in a short term. However, according to the evaluation hosted by the National Institute of Standards and Technology (NIST), Google's system was far ahead of peer systems. For example, in the large data track evaluation of Arabic-to-English tasks, Google system's BLEU score* was 51.31%, ahead of the run-ner-up system by almost 5%. However, improving this 5% in the past would

---

*A measurement system of machine translation quality. Generally speaking, the BLEU score of accurate artificial translation is 50%-60%.

TABLE 29.2: NIST evaluation results of many machine translation systems around the world (as of 2005).

Arabic-to-English,	Large data track
Google	51.31%
USC/ISI	46.57%
IBM	46.46%
UMD	44.97%
JHU-CU	43.48%
SYSTRAN	10.79%
Arabic-to-English,	Unlimited data track
Google	51.37%
SAKHR	34.03%
ARL	22.57%
Chinese-to-English,	Large data track
Google	35.31%
UMD	30.00%
JHU-CU	28.27%
IBM	25.71%
SYSTRAN	10.79%
Chinese-to-English,	Unlimited data track
Google	35.16%
ICT	12.93%
HIT	7.97%

require 5-10 years of research. In the unlimited data track evaluation, Google's BLEU score was 51.37%, ahead of the runner-up system by 17%, which is equivalent to leading by a generation. In Chinese-to-English translation, Google's advantage was equally apparent. Table 29.2 is the NIST evaluation results from 2005.*

People marveled at Google's achievement in machine translation and wondered how Och accomplished it. Although he is a world-class expert and had developed two great machine translation systems at the Technical University of Aachen in Germany and at the Information Sciences Institute (ISI) of the University of Southern California (USC), his time at Google was only enough to reimplement the old systems, and too short to conduct additional new research. According to NIST regulations, all participating organizations need to share their approaches after the evaluation results come out. Therefore, in July 2005, scholars from all over the world gathered at Florida for the conference hosted by NIST. Google's report was much anticipated because everyone was curious about its secret.

It turned out that Och's secret recipes were really simple when he revealed them. He used the same approach as two years before, but with tens of thousand

---

*Source: http://www.itl.nist.gov/iad/mig/tests/mt/2005/

folds of data other institutions used, and trained a six-gram language model. We mentioned in the previous chapters that an accurate estimate of each conditional probability (parameter) in an N-gram model requires sufficient data. The bigger N is, the more data we need. Generally speaking, N is no larger than 3. After 2000, some research institutions were able to train and use four-gram models, but that was all they could do. If the data size was doubled or tripled, the machine translation result may not be improved or just be marginally better. Using a data size 10 times the original, the result might improve by one percent. However, when Och used an enormous data volume that was tens of thousands times more than anyone else's, the accumulation of quantitative changes amounted to a qualitative transformation. It is worth mentioning that SYSTRAN is a professional machine translation company that uses grammar rules to implement translation. Before scientists knew how to use statistical methods in machine translation, it was a world leader in the field. However, compared to today's data-driven translation systems supported by statistical models, its product seems rather backward.

The second example is about speech recognition. According to Dr. Peter Norvig, my boss at Google, the error rate of speech recognition was reduced from about 40% in 1995 to about 20%-25% in 2005, and this significant improvement was largely due to the increase of training data. Dr. Norvig estimated that less than 30% of the credit should give to methodological improvements, while more than 70% went to data. In 2004, Google recruited the founding team of Nuance, a world-renowned speech recognition company. Dr. Michael Cohen, co-founder of Nuance, who later headed Google's speech recognition team, led a dozen scientists and engineers to develop a phone speech recognition system called Google-411. Compared to Nuance's original system, Google-411 did not have a higher recognition rate at first. By the standard of

FIGURE 29.3: Google's data center in North Carolina. A cabinet in this picture stores more data than all textual contents in the Library of Congress.

traditional research logic, this was completely meaningless. However, Google gathered enormous amounts of voice data through providing this free service to a large number of users, which prepared for Google's real speech recognition product Google Voice. With these data, Google was able to provide the most accurate speech recognition service to this date.

Since data is very useful, we will be able to make many discoveries out of it if we have more comprehensive data in larger quantities. The concept of big data was born in such circumstance.

29.3 Why we need big data

What is big data? Big data obviously has a large amount of data. However, a large data amount alone does not mean it is big data. The poll data that *The Literary Digest* collected was a lot, but it does not count as big data. The more important criteria are data's multidimensionality and its completeness. Only with these two features can we link seemingly unrelated events together, and reveal a comprehensive description of an object. In order to illustrate this, we will first look at a specific example.

In September 2013, Baidu published a rather interesting statistical result - *China's Top Ten "Foodie" Provinces and Cities*. Baidu did not conduct any surveys or research about local cuisines around China. It only "mined" some conclusions from 77 million food-related questions from "Baidu Zhidao."* However, these conclusions reflect the food preferences of different areas in China better than any academic research. Here are some of Baidu's conclusions:

For "is this edible" questions, the most common questions from netizens of Fujian, Zhejiang, Guangdong, and Sichuan provinces are some versions of "is this kind of insect edible"; netizens from Jiangsu, Shanghai, and Beijing frequently posted questions like, "is the skin of this edible"; netizens from Inner Mongolia, Xinjiang, and Tibet wondered "is mushroom edible"; and netizens from Ningxia asked "are crabs edible". The popular question from Ningxia (an inland province) is probably shocking to crab-loving people from Jiangsu and Zhejiang (coastline provinces). On the other hand, Ningxia netizens must be surprised to learn that there are people who want to eat insects!

This Baidu report is in fact a typical application of big data. It has the following characteristics: first, the data size itself is very big; 77 million questions and answers are a lot of data. Second, these data have many dimensions. They touch upon food preparation, ways to eat it, ingredients, nutrition information, price, the source location and time of questions asked, and so forth. Moreover, these dimensions are not explicitly provided (this is different from traditional databases). To an outsider, these raw data look very messy, but it is these

---

*Baidu's search-based Q&A forum.

"messy" data that connect the seemingly unrelated dimensions (time, location, food, preparation, ingredient, etc.) together. Through the mining, processing, and organizing of this information, Baidu was able to obtain meaningful statistical patterns such as the cuisine preferences of people living in different regions in China.

Of course, Baidu only published results that the public would be interested in. If asked, Baidu could present you even more valuable statistical results from these data. For example, it would be easy to obtain the eating habits of people of different ages, genders, and cultural backgrounds (suppose Baidu knows that the users' personal information is reliable; even if it is not complete from Zhidao service itself, Baidu obtains reliable personal information, like ages, or the eating habits, professions, even lifestyles (such as early risers, night owls, travelers, the sedentary, etc.) and so forth, of their users. Baidu's data was also collected over a long enough timespan, so it can even reveal the changes in different areas' eating habits, especially the changes at different stages of economic development. Without the big data of "Baidu Zhidao," it would be difficult to obtain reliable information regarding seemingly simple questions such as changes in eating habits.

Here, the reader might be wondering: the above statistical analysis does not seem too complicated to be obtained through traditional statistical means. I am not saying that traditional approaches do not work here, but they are much more difficult than what you might imagine. Let us think about what we would have to do in order to accomplish this through traditional means. First, we have to design a reasonable survey (which is not that easy), then we need to find representative people from different regions to distribute our survey to (which is what Gallup has been trying to do), and lastly, we need to process and organize the data semi-manually. This is not only costly, but also difficult to consider every factor comprehensively in the sampling process. This is also Gallup poll's problem. If we realize that there should be an additional item in the survey when we are counting the statistics data, it would be too late and costs twice as much.

The second reason that makes the traditional approach very difficult is that survey results do not always necessarily reflect the participant's actual opinions. We know that there is no pressure to ask or answer questions on "Baidu Zhidao," and there is no profit involved. People ask their own questions and answer what they know. However, the case is different when it comes to filling out surveys. Most people do not want to appear "weird" on the survey, so they might not truthfully report that they enjoy eating "smelly tofu" or "love eating insects." China Central Television has encountered a similar situation in the past when it was conducting surveys about what programs users watched. The self-reported results turned out to be completely different from the data automatically collected from the TV boxes. According to the self-reported data, people exaggerated how much they watched the high-quality shows hosted by famous anchors because it was natural to want to make themselves appear to have good taste. In the 2016 presidential election, candidate Hillary Clinton led in almost all surveys. However, data obtained from social networks showed her opponent, Donald Trump could win, and it turned out the social media was correct.

Besides lowering cost and high accuracy, big data also has the advantage of providing multidimensional information. In the past, computers had limited data storage and processing abilities, so only the data relevant to the problem to be solved was collected. These data had only a few dimensions, and those dimensions that appeared unrelated were neglected. This limitation determined a specific approach to use data, which was to propose a hypothesis or conclusion first and then to use data to prove it. Now, the emergence of cloud computing makes it possible for us to store and process large amounts of data with complex associations and even data that originally seemed irrelevant. The workflow also changes because of this. Besides using data to prove existing conclusions, we can start with the data without any preconceived notions to see what new conclusions the data itself may lead us to. This way, we can discover many new patterns. For example, the data in Wikipedia may appear to be very messy, but there are in fact many internal connections. Before analyzing these big data, product managers did not have any preconceived hypotheses in their minds or any presumed conclusions. However, after analyzing these data, many new patterns emerge. I suspect that these new patterns from big data would be shocking even for people working for Wikipedia.

Of course, there are many things in the world that are more important than eating insects or crabs, such as healthcare. We know that many diseases are related to genetic deficiencies, but genes work in complicated ways. A deficiency in one gene may cause a certain disease, but it is only a possibility. In order to figure out the connection between genes and diseases, the medical field generally has two approaches. The first approach is rather traditional. First, we need to understand the principle of a genetic sequence through experiment (this is a long process that usually has to start from fruit flies experiments), and understand the biological changes that can result from its deficiency. Then, we need to know whether such changes lead to diseases, or in what circumstances they lead to diseases. For example, suppose a certain genetic sequence is related to the synthesis of insulin. If it has problems, the metabolism of sugar may be affected, which could cause diabetes under certain conditions. Lastly, we would come up with this conclusion: "if a certain genetic sequence has a deficiency, it may cause diabetes." However, I used many "maybes" here, because all we know is that there is a possibility. As for how likely it is, nobody has the answer. The more difficult challenge is finding a causal relationship between the genes and the diseases. In short, this method takes a long time and a lot of money. For example, scientists around the world had been studying the issue for decades, but it was still difficult to pin down the causal relationships between smoking and many diseases. This enabled the big tobacco companies to avoid legal punishments until the late 1990s.

The second method uses data and statistics. These scientists' approach is the complete opposite of the first kind - they start with the data, find the statistical correlation between genetic deficiencies and diseases, and then analyze the internal factors causing such correlation. In our example of genetic deficiency and diabetes, if we only look for the relationship between a certain genetic sequence deficiency and the diabetes incidence rate, we can use a conditional probability to compute the likelihood of this genetic deficiency causing diabetes. The specific equation is as follows:

Suppose event A = {a certain genetic deficiency}, event B = {diabetes}, then the conditional probability is

$$P(B|A) = P(AB)/P(A) \approx \frac{\#(AB)}{\#(A)} \qquad (29.2)$$

Here, $\#$ represents the number of samples.

Although this appears to be very simple, it was not easy to implement in the past without big data.

First, when computing $\#$(AB), there are not too many cases that can establish a connection between a genetic deficiency and diabetes. In statistical terms, the data is too sparse. For example, in the US, few hospitals have thousands of diabetes patients, and it is possible that only 5% of these patients' genetic data have been extracted and stored in the hospital's database. Thus, we only have tens or dozens of people, and half of those people's diabetes could be unrelated to the genetic deficiency. Now, the remaining data is too little for obtaining any reliable statistical pattern.

Second, the denominator in Equation (29.2), $P(A)$, is impossible to obtain. This is because we can never know for sure how many people have this genetic deficiency. Before the era of big data, this seemingly simple task was in fact impossible to accomplish. In order to sue tobacco companies, the US sent some experts to China to collect data in the late 1990s because all of American smokers' data combined was not enough for a reliable statistical result.

The other challenge in researching the relationship between the human genome and diseases is how to find those potentially defective genes. We need to know that the complete genetic data of a human is enormous. According to Dr. Huanming Yang, co-founder of Beijing Genomics Institute (BGI), the amount of data is unimaginably large, at the degree of magnitude of PB (10^{15} bits, i.e., one million GB) (Huanming Yang, 2013, personal communication).* Considering the data size, the genetic data for one person has already exceeded the amount of data of "Baidu Zhidao." Of course, if we only look at one person's genes, there is no way to tell which sequence is good or which one is deficient. Even if we collect several people or dozens of people, the genes would not be enough to tell us anything. This is because there are individual genetic differences, which means that the difference in genes does not necessarily indicate a genetic deficiency. In order to pinpoint potential deficiencies, we need the genetic data of at least tens of thousands of people. Before cloud computing, it was impossible for people to process such an enormous amount of data.

---

*This includes the genetic data of main bacteria in the human body, because they affect a person's health and are related to diseases.

Collecting genetic data from a large number of people used to be a challenge as well. Fortunately, things that appear to be very difficult always turn out to have solutions. There is a small American company called 23andMe. Its product is very interesting and smartly designed. This company only charges $100 (not the kind of comprehensive genetic screening at hospitals that require $1000-$5000) to collect your saliva, then it "reads" your genes roughly, and tells you "generally" the probabilities of you having various diseases in the future. Of course, this company's genetic reading is not the same thing as the genomic mapping that BGI does. Nonetheless, $100 is insufficient for even a rough genetic analysis. 23andMe in fact uses this method to attract a large number of genetic information providers. With a large amount of genetic data, they can distinguish which genetic fragment is normal, and which has a "potential" deficit. For each genetic information provider, they can enumerate this person's potential deficient genes. Of course, they can also obtain the probability of each kind of genetic deficiency, which is $P(A)$ in the above Equation (29.2).

23andMe and similar companies (including Calico, Human Longevity Inc., etc.) are also trying to make connections between genetic deficiencies and diseases. This data must be obtained from research institutions and hospitals. In the past, each hospital only had limited data in this respect, but if we could gather data from tens of thousands of hospitals large and small,* then we can estimate the probability of a disease and a genetic deficiency co-occurring, and thus compute the probability of a certain genetic deficiency causing a disease. In the future, big data can provide accurate predictions of each person's future health using genetic screening, which will facilitate disease prevention.

I used a healthcare example because, besides the IT industry, the medical field is most excited about big data. Of course, the other reason is that both Google and I are interested in healthcare, which makes it easy to come up with an example. However, this does not mean that big data is only applied in these two industries.

The healthcare industry is the largest industry in the US. In 2017, its output value was about 18% of the total GDP. If the cost does not decrease, it would take up about 20% of GDP. Although doctors have long dealt with data (all kinds of testing results) every day in this big industry, they were not too motivated to use information technologies to improve the quality of care in the past fifty to sixty years (except for using imaging technologies). However, this situation was changed in the past decade. The healthcare industry took the initiative to make contact with the IT industry in order to solve difficult healthcare problems with big data. This also shows us the importance of big data. Up-to-date, big data has brought many unexpected improvements to the healthcare

---

*Some people might be wondering why these hospitals are willing to provide such data to a third party. This is because they realize that the data they have is not enough to make any convincing conclusions, whereas large integrated hospitals, like Kaiser Permanente, are unwilling to provide data since they can conduct the research themselves.

industry. In 2012, US media reported on two representative applications of big data in the medical field.

The first story is about a high school student named Brittany Wenger. In 2012, she made a significant improvement in the accuracy of breast biopsy location using big data. We know that patients who are at risk for breast cancer need to undergo biopsy, which is using a special needle to extract some cells at the target area to test for cancer cells. The accuracy of biopsy depends entirely on the puncture location. If the location is inaccurate, then the cancer cells will not be detected even if they exist. In the past, the accuracy depended on the doctor's experience and expertise. However, in the US, a doctor might only encounter a few hundred such cases in an entire career, which means that the experience accumulation process is very slow. Even for experienced doctors, it is difficult to maintain stable performance every day (because the doctor's mood changes will affect the accuracy of judgment). So what did this high school student do? She gathered millions of cases and wrote a program that circles suspicious target areas on X-ray and CT images, the accuracy of which reached 98%. This is much higher than the traditional experience-based approach. Of course, the millions of cases that she used do not seem to be a large amount of data to people who work in the information processing field, but it is very big for the healthcare industry. Her research result won first place at the 2012 Google Science Fair competition that year.*

The second story is about Microsoft big data application.† Insurance companies notice that many emergency patients are transported back to the emergency room shortly after being discharged from the hospital. Since emergency care costs a lot, it is a considerable burden for individuals as well as for insurance companies. Thus, insurance companies and hospitals gathered large amounts of patient information and gave it to big data scientists and engineers at Microsoft to analyze for contributing factors or statistical patterns. Using machine learning algorithms, Microsoft staff extracted and analyzed tens of thousands of features. Eventually, they found that if a patient was put on a drip, the likelihood of returning to the ER in several weeks would be very large (probably because the condition is very serious). Other important features were also found in addition to this pattern. Therefore, for patients with these features, as long as they are followed up on and checked periodically following discharge from the hospital, their likelihood of returning to the ER would be greatly reduced. Thus, the healthcare fees would also decrease significantly. Training a mathematical model with tens of thousands of features is impossible without multidimensional big data.

More research that use big data to support healthcare is conducted in university and company labs, including a very meaningful study that uses big data to

---

*http://www.huffingtonpost.com/2012/07/25/brittany-wenger-google-science-fair-breast-cancer_n_1703012.html

†http://www.microsoft.com/casestudies/Microsoft-Azure/Leeds-Teaching-Hospital/Big-Data-Solution-Transforms-Healthcare-with-Faster-Access-to-Information/710000003776

prescribe medicines for different symptoms. At the Stanford computational biology research center, some professors and students use big data to conduct pairing studies with thousands of medicines and symptoms. They found that a heart disease medicine turned out to be highly effective for relieving some patients' gastrointestinal symptoms. Through this research, they found many new treatments for various illnesses at a much lower cost and in a shorter amount of time than developing new drugs.

Lastly, I will tell you something very exciting. This thing could benefit all of us using big data. I will use it as the end of this chapter and of this book.

In 2013, Google founded Calico, a company that uses IT achievements to solve healthcare problems, and hired world-famous biopharmaceutical expert and former Genentech CEO, Dr. Arthur D. Levinson, to lead this creative initiative. As the chairman of Apple and Genentech, why would Dr. Levinson and many of his past colleagues go to a tech company with no healthcare or biopharmaceutical experience to do healthcare research? This was because they knew that the future would be ruled by data. Many challenges, such as curing cancer and preventing aging, are impossible to conquer with traditional medical approaches. In order to solve these problems, we need technologies related to big data.

At a presentation, Dr. Levinson explained why humans still could not cure cancer today. He thought that there were two main reasons. First, the effectiveness of a drug is closely related to a person's genes. Thus, we need to prescribe different drugs to people with different genes. Taking this to an extreme would be designing a special kind of drug for each individual. However, even if this method works, we have to consider the cost problem. According to his estimates, inventing a new cancer drug for someone using conventional methods for drug development would cost one billion dollars per patient, which is of course too high to be generalized. Second, the genes of cancer cells themselves are constantly changing. We often hear about cases where the patient improves quickly at first when using a new cancer drug, but the drug ceases to be effective later on and the cancer relapses uncontrollably. This is because the genes of cancer cells have already changed, so the drug that worked on the original cells no longer has any effect. According to Dr. Levinson, the biggest problem at present is that even if we develop a special cancer drug for everyone, the current drug developing speed is not as fast as the change of cancer cells. Dr. Levinson thought that in order to solve these two problems (cancer drugs have to be individualized and drug development has to be faster than cell genetic changes), we have to rely on big data. Statistics for similarities across individuals can prevent repetition in experiments for new drug development, and only limited animal testing needs to be done before conducting clinical trials. Eventually, the cost for designing personalized drugs for each person would be limited to within $5000. At the same time, since most of the work can be shared, the drug update cycle can be shortened so that new drugs are developed faster than the cancer cell changes, giving us hope for curing cancer.

Now, Dr. Levinson and colleagues are using Google's platform to integrate healthcare resources all over the US in order to solve the longevity problem, which has concerned people around the world for thousands of years. We hope that he can bring us the good news.

From the above examples, we can see that big data has a large impact on the information industry as well as other fields. Now, let us summarize the importance of big data. First, meaningful statistical patterns can only be obtained when the combinations of random events co-occur many times. Second, the gathering of big data is a natural process, which helps eliminate subjective bias. Of course, more importantly, only multidimensional big data can make seemingly loosely connected events occur repeatedly and discover new patterns. Lastly, big data may be the key to solving many hard problems outside of the IT industry (for example, in the healthcare industry).

29.4   Summary

Although people have understood the importance of data for a very long time, they were satisfied with "enough" data due to storing and computing limitations in the past. As information technology developed, storage and computation were no longer problems. People started to realize that huge amounts of data brought unexpected findings, which led to the rise of big data.

In the future, data will become increasingly indispensable in people's life. Many job opportunities surrounding data collection and processing will continue to emerge, and experts in processing and using data will become the successful people in the new era. Generally speaking, regardless of the field or the job, whoever understands the importance of data and knows how to use data well will be more likely to succeed in the future world.

Bibliography

1. Ju Sheng, Shiqian Xie, Chengyi Pan. "Probability Theory and Mathematical Statistics (Gai lü lun yu shu li tong ji)." 4th edition. Gao deng jiao yu chu ban she, 2010.

2. Thorsten Brants, Ashok C. Popat, Peng Xu, Franz J. Och, Jeffrey Dean. "DeViSE: A Deep Visual-Semantic Embedding Model." Large Language Models in Machine Translation. Proceedings of the 2007 EMNLP-CoNLL, pp.858–867.

Postscript

Many friends ask what made me decided to write the blog series *The Beauty of Mathematics* in the Google China Official Blog. Although the original purposes that the company published official blogs were to introduce Google's technologies and products directly to Chinese users, I always believed that users would like your products better when they understood the background technologies behind feature they could see. As a research scientist who used mathematics in my daily work at Google, I tried to explain the mathematical foundation of Google products in a language that most people could easily understand. I also liked to should young engineers how to use mathematics they learned in college in their engineering works. In the US as well as in China, I often see that the majority of software engineers rely on their intuitive feelings to navigate unknown territories, solving problems using makeshift methods. This problem is especially pronounced in China. Such an approach is highly unsystematic and unprofessional. When I first joined Google, I noticed that some early algorithms (such as spellcheck) had no systematic algorithms or theoretical foundations, but were entirely made up of makeshift phrases or doubles of words. Although these methods are better than nothing, they have no potential to progress or improve. Instead, they make the program's logic rather chaotic. As the company expanded and grew, Google started to hire engineers with excellent theoretical backgrounds from the world's best universities, which ensured engineering projects' correctness. In the following year, I mentored several graduate students from some of the world's best universities. Using a hidden Markov model framework, we unified all of Google's spellcheck models. During those years, Google rewrote almost all of the projects' programs, and there were no more makeshift methods. However, in other companies, including some second-rate IT companies that are ostensibly titled high-tech, bad practices were still prevalent. In China, some small startup companies sacrifice quality for quantity, which is understandable. However, makeshift approaches persist in some listed companies with profits that rank high in the world, giving people a rather poor impression of them. On the other hand, these companies often surpass many transnational corporations in their spendings on office buildings and management offices. This is like the nouveau-riche who dresses up fancy but neglects to make progress in knowledge and character. Hence, I wrote *The Beauty of Mathematics in Computer Science* in the hope that these companies' engineering supervisors can lead their teams to improve the engineering quality and move away from poor practices. Only this way can they approach the level of world-class IT companies and avoid astonishing waste of resources due to large amounts of low-level repetitive constructions.

Wrong models (adopted unintentionally) might suffice temporarily for specific occasions, such as our geocentric model - people had used it for thousands of years after all. However, wrong models stray from the truth, and their negative influences would eventually emerge. As a result, a wrong model not only leads us further away from the truth, but often complicates something that could have been simple to the verge of destruction (this is what the geocentric model did to the heliocentric model).

The process to discover correct theories and methods is gradual. Everything has its pattern of development, and these patterns can be discovered. The information technology field is no exclusion. Once we discover patterns, we should consciously follow them in our work and not go against them. Dr. Shannon was someone who discovered the developmental pattern of information science. His information theory points out the nature and patterns of information processing and communication today to a large extent. Here, communication covers all kinds of exchanges of humankind, including all applications of natural language processing. When I first started writing this blog series, introducing these patterns of information processing was what I had in mind.

Of course, it is not an easy task to explain mathematical concepts clearly so that even outsiders to the field can understand. I used to think that I was rather good at illustrating technical principles using simple language, but when I first presented several chapters to readers without any engineering background, they told me that it was difficult for them to grasp. Later, I spent a lot of time and effort to try to make this series understandable and approachable to everyone. However, this necessitated sacrificing many details, which upset some professional readers who wished that I would introduce more technical details. After leaving Google, there were less constraints to my writing, so I added some more technical details when publishing the blog series in this book format for those readers with strong engineering backgrounds who are willing to learn more about details. When I was finished with the book, I realized that compared with the original blog series, the whole book's content was practically rewritten. For this new edition, I added some new content as well as additional details, which I hope can be helpful to readers at all levels.

One of my main objectives for writing the book is to introduce some mathematical knowledge related to the IT industry to people who do not work in the industry. Hence, I hope that this book could be a popular science book that people read for leisure. Through understanding patterns of information technology, the readers can apply what they learn to summarize and use the patterns in their own work, and thus work toward self-improvement.

Two books and one TV series helped me the most in writing this book. When I was in middle school, I read a popular science book by Russian-American physicist, George Gamow, called *One Two Three... Infinity: Facts and Speculations of Science* that introduced the fundamentals of the universe. He spent a lot of time writing popular science books that influenced generations of readers. The second book was by the famous British physicist, Stephen Hawking: *A Brief History Of Time*. Hawking used simple language to explain profound principles of the universe, which made this popular science book a bestseller around the world. A TV series that had a big impact on me was *Through the Wormhole*, hosted and narrated by the famous actor, Morgan Freeman. Most of my writing is completed on planes, and I would take breaks to watch some TV during periods of writing. Once, I stumbled upon *Through the Wormhole*, a series that makes the most up-to-date physics comprehensible. Many world-class physicists and mathematicians, including Nobel Prize winners, come to the show to introduce their work. These people have a common skill, which is the ability to explain the deepest principles in their fields using simple metaphors so that the general public can understand. I think this may be one of the reasons that they became top scientists: they are very knowledgeable in their fields and at the same time good at explaining things in plain words. Good scholars always find ways to explain complex concepts in simple terms to outsiders, instead of complicating simple problems to put on an aura of mystery. Therefore, when writing *The Beauty of Mathematics*, I had aspired to the models of Gamow and Hawking, and tried to demonstrate the beauty of mathematics to all readers rather than only to those with relevant technical backgrounds. In order to make it convenient for readers to enjoy the book during small chunks of leisure time, I tried to make each chapter relatively independent and self-contained in my writing. After all, it is difficult to ask most readers to read a book about mathematics from cover to cover, all in one sitting.

Jun Wu
October 2014, Silicon Valley

Index

Printed in the United States
by Baker & Taylor Publisher Services